PIECEWISE LINEAR STRUCTURES ON TOPOLOGICAL MANIFOLDS

PIECEWISE LINEAR STRUCTURES ON TOPOLOGICAL MANIFOLDS

Yuli Rudyak

University of Florida, USA

World Scientific

NEW JERSEY · LONDON · SINGAPORE · BEIJING · SHANGHAI · HONG KONG · TAIPEI · CHENNAI · TOKYO

Published by

World Scientific Publishing Co. Pte. Ltd.

5 Toh Tuck Link, Singapore 596224

USA office: 27 Warren Street, Suite 401-402, Hackensack, NJ 07601

UK office: 57 Shelton Street, Covent Garden, London WC2H 9HE

Library of Congress Cataloging-in-Publication Data

Names: Rudyak, Yuli B., 1948–

Title: Piecewise linear structures on topological manifolds / by Yuli Rudyak
   (University of Florida, USA).

Description: New Jersey : World Scientific, 2016. | Includes bibliographical references and index.

Identifiers: LCCN 2015037499 | ISBN 9789814733786 (hardcover : alk. paper)

Subjects: LCSH: Topological manifolds. | Manifolds (Mathematics) | Piecewise linear topology.

Classification: LCC QA613.2 .R83 2016 | DDC 514/.34--dc23

LC record available at http://lccn.loc.gov/2015037499

British Library Cataloguing-in-Publication Data

A catalogue record for this book is available from the British Library.

Printed in Singapore

Dedicated to Irina and Marina

Preface

The study of triangulations of topological spaces has always been at the root of geometric topology. Among the most studied triangulations are piecewise linear triangulations of high-dimensional topological manifolds. Their study culminated in the late 1960s–early 1970s in a complete classification in the work of Kirby and Siebenmann. It is this classification that we discuss in this book, including the celebrated Hauptvermutung and Triangulation Conjecture.

The goal of this book is to provide a readable and well-organized exposition of the subject, which would be suitable for advanced graduate students in topology. An exposition like this is currently lacking. The foundational monograph of Kirby and Siebenmann [KS2] proving the classification was written on the heels of the proof. It contains all the necessary ingredients but, written in the form of essays, can hardly serve as an exposition. Another very useful source of information on the subject, the book of Ranicki [Ran], has the same drawback as being a collection of research papers.

In this book, I attempted to give a panoramic view of the theory. Given in how many different directions this theory branches out, I took special care not to lose sight of the forest for the trees. For instance, I chose to merely state several well-known theorems, providing references to well-written proofs available in the literature.

Acknowledgments. The work was partially supported by Max-Planck of Mathematics, Bonn, and by a grant from the Simons Foundation (#209424 to Yuli Rudyak). I express my best thanks to Andrew Ranicki who read the first version of the manuscript and did many useful remarks and comments. I am grateful to Hans-Joachim Baues, Moris Hirsch, Tom Farrell, and Nikolai Saveliev for useful discussions. I am also grateful to Rochelle Kronzek, E. H. Chionh and Rajesh Babu of World Scientific for the help in preparing the manuscript for publication.

Yuli Rudyak, *Gainesville, Florida*, November 2015

Contents

Introduction

Throughout this volume we use abbreviation PL for "piecewise linear".

For introduction to PL topology, including definitions of simplicial complexes, polyhedra, PL maps, etc. see [Hud, RS]

Hauptvermutung (main conjecture) is an abbreviation for *die Hauptvermutung der kombinatorischen Topologie* (the main conjecture of combinatorial topology). It seems that the conjecture was formulated in the papers of Steinitz [Ste] and Tietze [Ti] in 1908. This is also stated in [AH].

The conjecture states that the topology of a simplicial complex determines completely its combinatorial structure. In other words, two simplicial complexes are simplicially isomorphic whenever they are homeomorphic. This conjecture was disproved by Milnor [Mi2] in 1961. In fact, Milnor found a pair of homeomorphic simplicial complexes such that the Whitehead torsion of this pair is non-trivial. See Cohen [C] for a textbook account.

Note, however, that the Whitehead torsion cannot distinguish homeomorphic *manifolds*, [KS1, C]. Thus, in case of manifolds, one can propose a refined version of the *Hauptvermutung* by considering simplicial complexes with natural additional restrictions. A *combinatorial triangulation* is defined to be a simplicial complex such that the star of every point (the union of all closed simplexes containing the point) is simplicially isomorphic to the n-dimensional ball. A *PL manifold*, or *combinatorial manifold* is defined to be a topological manifold M together with a homeomorphism $M \to K$ where K is a combinatorial triangulation. Equivalently, a PL manifold can also be defined as a manifold equipped with a maximal PL atlas.

There exist topological manifolds that are homeomorphic to a simplicial complex but do not admit a PL structure (non-combinatorial triangulations), see Example 3.5.5. Furthermore, there exist topological manifolds that are not homeomorphic to any simplicial complex, see Example 3.5.7.

Now, the *Hauptvermutung for manifolds* is the conjecture that any two homeomorphic PL manifolds are PL homeomorphic. The related *Combinatorial Triangulation Conjecture* states that every topological manifold admits a PL structure, i.e., can be triangulated by a PL manifold. Both these conjectures were disproved by Kirby and Siebenmann [Sieb4, KS1, KS2]. In fact, Kirby and Siebenmann classified PL structures on high-dimensional ($\geqslant 5$) topological manifolds. It turned out that a topological manifold can have different (not PL homeomorphic, non-concordant) PL structures, as well as having no PL structures. Now we give a brief description of these results.

In dimensionals $\leqslant 3$ every topological manifold admits a PL structure that is unique up to PL homeomorphism, see [Rad, P, Mo]. The classifcation of PL structures on 4-dimenaional topological manifolds is not completed yet, cf. [FQ, K2].

Let $BTOP$ and BPL be the classifying spaces for stable topological and PL (micro)bundles, respectively. We regard the forgetful map

$$\alpha : BPL \to BTOP$$

as a fibration and denote its homotopy fiber by TOP/PL.

Let $f : M \to BTOP$ classify the stable tangent bundle of a topological manifold M. By the main properties of classifying spaces, every PL structure on M gives us an α-lifting of f and that every two such liftings for the same PL structure are fiberwise homotopic.

It is remarkable that the converse is also true if $\dim M \geqslant 5$, see [LR1, KS2]. In greater detail, M admits a PL structure if f admits an α-lifting (*the Existence Theorem* 1.7.4), and concordance classes of PL structures on M are in a bijective correspondence with fiberwise homotopy classes of α-liftings of f (*the Classification Theorem* 1.7.2). So, the homotopy information on the space TOP/PL is extremely useful in PL classifying of topological manifolds. Fortunately, Kirby and Siebenmann have made great progress there: they proved the following

Main Theorem: *There is a homotopy equivalence*

$$TOP/PL \simeq K(\mathbb{Z}/2, 3).$$

Thus, there is at most one possible obstruction

$$\varkappa(M) \in H^4(M; \pi_3(TOP/PL)) = H^4(M; \mathbb{Z}/2)$$

to an α-lifting of the map f.

In particular, a topological manifold M, $\dim M \geqslant 5$ admits a PL structure if and only if $\varkappa(M) = 0$. Furthermore, the set of fiberwise homotopic α-liftings of f (if they exist) is in a bijective correspondence with $H^3(M; \mathbb{Z}/2)$. At manifolds level, we can say that every homeomorphism $h : V \to M$ of a PL manifold V yields a class

$$\varkappa(h) \in H^3(M; \mathbb{Z}/2),$$

and $\varkappa(h) = 0$ if and only if h is concordant to the identity map 1_M. Moreover, every class $a \in H^3(M; \mathbb{Z}/2)$ has the form $a = \varkappa(h)$ for some homeomorphism $h : V \to M$ of two PL manifolds.

These results yield the complete classification of PL structures on a topological manifold of dimension $\geqslant 5$. In particular, the situation with *Hauptvermutung* turns out to be understandable. See Section 3.4 for more detailed exposition.

We would like to explain the following. It can happen that non-concordant PL structures on M yield PL homeomorphic PL manifolds (like that two p-liftings $f_1, f_2 : M \to BPL$ of f can be non-fiberwise homotopic). Indeed, a PL map $M \to M$ of a PL manifold M can turn the atlas into a non-concordant to the original one, see Example 3.5.3. So, in fact, the set of pairwise non-concordant PL manifolds which are homeomorphic to a given PL manifold is in a bijective correspondence with the set $H^3(M; \mathbb{Z}/2)/R$ where R is the following equivalence relation: two concordance classes of PL structures are equivalent if the corresponding PL manifolds are PL homeomorphic. The *Hauptvermutung* for manifolds states that the set $H^3(M; \mathbb{Z}/2)/R$ is a singleton for all M. But this is wrong in general.

Namely, there exists a PL manifold M which is homeomorphic but not PL isomorphic to $\mathbb{R}P^n$, $n \geqslant 5$, see Example 3.5.1. So, here we have a counterexample to the *Hauptvermutung*.

To complete the picture, we mention again that there are topological manifolds that do not admit any PL structure, see Example 3.5.4. Moreover, there are manifold that cannot be triangulated as simplicial complexes, see Example 3.5.7.

Recall that every smooth manifold admits a canonical PL structure [Cai, W], while every PL manifold is, tautologically, a topological
manifolds. Now we compare the classes of smooth, PL and topological
manifolds, and see that there is a big difference between first and second
classes, and not so big difference between second and third ones. From
the homotopy-theoretical point of view, one can say that the space PL/O
(which classifies smooth structures on PL manifold, see Remark 1.7.8)
has many non-trivial homotopy groups, while the space TOP/PL is an
Eilenberg–MacLane space. Geometrically, one can mention that there are
many smooth manifolds which are PL homeomorphic to standard sphere
S^n but pairwise non-diffeomorphic [KM], while any PL manifold $M^n, n \geqslant 5$
is PL homeomorphic to S^n provided that M is homeomorphic to S^n, [Sma].

It is interesting and worthwhile to go one step deeper and explain the
following. Recall that a manifold M is called *almost parallelizable* if M
becomes parallelizable after deletion of a point. Let σ_k^S (resp. σ_k^{PL}, resp.
σ_k^{TOP}) denote the minimal positive integer number which can be realized
as the signature of the closed smooth (resp. PL, resp. topological) almost
parallelizable $4k$-dimensional manifold. Clearly, $\sigma_k^S \geqslant \sigma_k^{PL} \geqslant \sigma_k^{TOP}$.

Let B_m denote the mth Bernoulli numbers, see [Wash] (we use the even
index notation, i.e., $B_{2n+1} = 0$). It turns out to be that

$$\sigma_1^S = 16 \text{ and } \sigma_k^S = 2^{2k+1}(2^{2k-1} - 1) \text{ numerator } (4B_{2k}/k) \text{ for } k > 1.$$

See [Ro] for $k = 1$ and [MK] or [MS, Appendix B] for $k > 1$. In particular,
σ_k^S strictly increases with respect to k.

Concerning the numbers σ^{PL} and σ^{TOP}, it turns out to be that

$$\sigma_1^{PL} = 16 \text{ and } \sigma_k^{PL} = 8 \text{ for all } k > 1,$$

and

$$\sigma_k^{TOP} = 8 \text{ for all } k.$$

First, for all k the number 8 divides the number σ_k by purely algebraic reasons, [Br2,
Proposition III.1.4]. Furthermore, $\sigma_1^{PL} = \sigma_1^S = 16$ since there is no difference between
PL and smooth cases up to dimension 6, see 1.7.8. Let W^{4k} be a $4k$-dimensional smooth
manifold with boundary (Milnor's pumbing) described in [Br2, Theorem V.2.1]. This is
a parallelizable manifold of signature 8. Furthermore, for $k > 1$ the boundary ∂W^{4k} is
a homotopy sphere. Hence, ∂W^{4k} is PL homeomorphic to the standard sphere by the
Smale Theorem [Sma]. So, the cone $C = C(\partial W^{4k})$ is PL homeomorphic to the standard
disk, and we get a closed almost parallelizable PL manifold $W^{4k} \cup_{\partial W^{4k}} C$ of signature
8.

To prove that $\sigma_1^{TOP} = 8$, consider the plumbing $W = W^4$ as above. Now
its boundary ∂W is not simply-connected, but it is a homology 3-sphere. Freedman

[F, Theorem 1.4′] proved that ∂W bounds a contractible topological 4-manifold P (in fact, this holds for any homology 3-sphere). Now, the the space

$$W \cup_{\partial W} P$$

is a closed almost parallelizable topology manifold of signature 8.

So, $\sigma_k^{TOP} = \sigma_k^{PL} > \sigma_k^{S}$ for $k > 1$, and we see again that there is a big difference between smooth and PL cases and not so big difference between PL and topological cases. Nevertheless, the last difference does not vanish, and the numerical inequality

$$16 = \sigma_1^{PL} \neq \sigma_1^{TOP} = 8$$

occurs whenever we meet a contrast between PL and topological world. For example, we will see below that the number

$$2 = 16/8 = \sigma_1^{PL}/\sigma_1^{TOP}$$

is another guise of the number

$$2 = \text{ the order of the group } \pi_3(TOP/PL).$$

In this context, it makes sense to notice about low-dimensional manifolds, because of the following remarkable contrast. There is no difference between PL and smooth manifolds in dimension < 7: every PL manifold $V^n, n < 7$ admits a smooth structure that is unique up to diffeomorphism. However, there are infinitely many smooth manifolds which are homeomorphic to $\mathbb{R}^4$ but pairwise non-diffeomorphic, see Section 3.5, Summary.

Concerning the description of the homotopy type of TOP/PL, we have the following. Because of the Classification Theorem, if $k + n \geqslant 5$ then the group $\pi_n(TOP/PL)$ is in a bijective correspondence with the set of concordance classes of PL structures on $\mathbb{R}^k \times S^n$. However, this set (of concordance classes) looks wild and uncontrollable. In order to make the situation more manageable, we consider PL structures on the *compact* manifold $T^n \times S^k$ and then extract the necessary information about the universal covering $\mathbb{R}^n \times S^k$ from here. We can't do it directly, but there is a trick (the Reduction Theorem 1.9.7 that is based by ideas of Kirby) which allows us to estimate PL structures on $\mathbb{R}^n \times S^k$ in terms of the so-called *homotopy PL structures* on $T^n \times S^k$ (more precisely, we should consider the homotopy PL structures on $T^n \times D^k$ modulo the boundary), see Section 1.4 for the definitions. Now, using results of Hsiang and Shaneson [HS] or Wall [W3, W4] about homotopy PL structures on $T^n \times D^k$, one can prove that $\pi_i(TOP/PL) = 0$ for $i \neq 3$ and that $\pi_3(TOP/PL)$ has at most 2 elements. Finally, there exists a high-dimensional topological manifold which does not admit any PL structure (Corollary 1.8.4, Remark 1.8.6). Hence, by the Existence Theorem, the space TOP/PL is not contractible. Thus, $TOP/PL \simeq K(\mathbb{Z}/2, 3)$.

For better arrangement of the previous matter, look at the graph located after the Introduction. We formulate without proofs the boxed statements (and provide the necessary preliminaries and references) there, while in Chapter 1 we explain how a statement (box) can be deduced from others, accordingly with the arrows in the graph.

Let me comment the top box of the graph. Sullivan [Sul1, Sul2] proved that the *Hauptvermutung* holds for simply-connected closed manifolds $M, \dim M \geqslant 5$ with $H_3(M)$ 2-torsion free.

In greater detail, let G_n be the monoid of homotopy self-equivalences $S^{n-1} \to S^{n-1}$, let BG_n be the classifying space for G_n, and let $BG = \lim_{n\to\infty} BG_n$. There is an obvious forgetful map $BPL \to BG$ (delete zero section), and we denote the homotopy fiber of this map by G/PL. For every homotopy equivalence of closed PL manifolds $h : V \to M$, Sullivan defined the *normal invariant* of h to be a certain homotopy class $j_G(h) \in [M, G/PL]$, see Section 1.5.

Let $M, \dim \geqslant 5$ be a closed PL manifold such that $H_3(M)$ is 2-torsion free. Sullivan proved that, for every *homeomorphism* $h : V \to M$, we have $j_G(h) = 0$. Moreover, this theorem implies that if, in addition, M is simply-connected then h is homotopic to a PL homeomorphism. So, as we already noted, the *Hauptvermutung* holds for simply-connected closed manifolds $M, \dim M \geqslant 5$ with $H_3(M)$ 2-torsion free.

Definitely, the above-mentioned Sullivan Theorem on the Normal Invariant of a Homeomorphism is important by itself. However, here this theorem plays also an additional substantial role. Namely, the Sullivan Theorem for $T^n \times S^k$ is a lemma in classifying of homotopy structures on $T^n \times D^k$, cf. Section 1.6. For this reason we first prove the Sullivan Theorem for $T^n \times S^k$ (the top box), then use it in the proof of the Main Theorem, and then (in Chapter 2) use the Main Theorem in order to prove the Sullivan Theorem in full generality.

I decided to present a proof of the Sullivan Theorem in the volume, Section 3.3 because the exposition in [Sul2] is quite intricate.

This volume is organized as follows. After the Introduction we present the above-mentioned graph, and the extensive comments of the graph appear in the first chapter. In other words, Chapter 1 contains the architecture of the proof of the Main Theorem.

The second chapter contains a proof of the Sullivan Theorem on the triviality of normal invariant of a homeomorphism for $T^n \times S^k$, i.e., we attend the top box of the graph.

The third chapter contains some applications of the Main Theorem. We complete the proof of the Sullivan Theorem on the triviality of the normal invariant of a homeomorphism in full generality. Then we tell more on classification of PL manifolds and, in particular, on Hauptvermutung. Several interesting examples are considered. Finally, we discuss the homotopy and topological invariance of certain characteristic classes.

Graph

Here I present the graph, and after the picture I list the boxed claims, with extraction the correspondent tags inside the body of the manuscript. Some minor comments are given. Some theorems here are stated simpler than those in the main text.

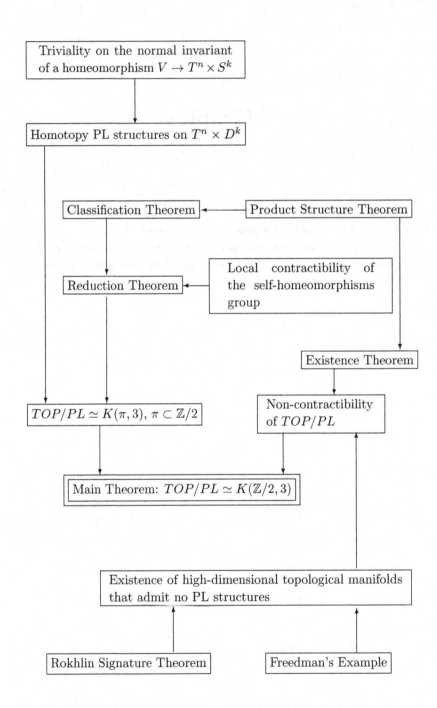

Graph xxi

1. Triviality of the normal invariant of a homeomorphism

$$V \to T^n \times S^k.$$

This is Theorem 2.8.1: If an element $x \in \mathcal{S}_{PL}(T^n \times S^k)$ can be represented by a homeomorphism $h : V \to T^k \times S^n$, then $j_G(x) = 0$.

Here and below we denote by $\mathcal{S}_{PL}(X)$ the set of equivalence classes of homotopy triangulations of a topological manifold X.

2. Homotopy PL structures on $T^n \times D^k$. This is Theorem 1.6.3: Supposed that $k + n \geqslant 5$. Then the following holds:

(i) if $k > 3$ then the set $\mathcal{S}_{PL}(T^n \times D^k)$ consists of precisely one (trivial) element;

(ii) if $k < 3$ then every element of $\mathcal{S}_{PL}(T^n \times D^k)$ can be finitely covered by the trivial element;

(iii) the set $\mathcal{S}_{PL}(T^n \times D^3)$ contains at most one element which cannot be finitely covered by the trivial element.

3. Classification Theorem. This is Theorem 1.7.2: If $\dim M \geqslant 5$ and M admits a PL structure, then the map

$$j_{TOP} : \mathcal{T}_{PL}(M) \to [M, TOP/PL]$$

is a bijection.

Here and below we denote by $\mathcal{T}_{PL}(X)$ the set of concordance classes of PL structures on a topological manifold X. The map j_{TOP} defined in 1.5.1.

4. Product Structure Theorem. This is Theorem 1.7.1: For every $n \geqslant 5$ and every $k \geqslant 0$, the map

$$e : \mathcal{T}_{PL}(M) \to \mathcal{T}_{PL}(M \times \mathbb{R}^k)$$

is a bijection. Here the map e turns a PL strucrture on M into a PL structure on $M \times \mathbb{R}^k$ in an obvious way: the product with $\mathbb{R}^k$. Roughly speaking, this theorem establishes a bijection between (the concordance classes of) PL structures on M and $M \times \mathbb{R}^k$. The Classification Theorem 1.7.2 and the Existence Theorem 1.7.4 are consequences of the Product Structure Theorem.

5. Reduction Theorem. This is Theorem 1.9.7. It reduces an evaluation of groups $\pi_i(TOP/PL)$ to the evaluation of sets $\mathcal{S}_{PL}(T^n \times D^k)$.

6. Local contractibility of the homeomorphism group. This is Theorem 1.9.1: The space of self-homeomorphisms of a compact manifold M is locally contractible.

7. $TOP/PL \simeq K(\pi, 3)$, $\pi \subset \mathbb{Z}/2$. This is Theorem 1.9.8.

8. Existence Theorem. This is Theorem 1.7.4: A topological manifold M with $\dim M \geqslant 5$ admits a PL structure if and only if the tangent bundle of M admits a PL structure.

9. Main Theorem: $TOP/PL \simeq K(\mathbb{Z}/2, 3)$. This is Theorem 1.9.9.

10. Non-contractibility of TOP/PL. This is Corollary 1.8.5.

11. Existence of high-dimensional topological manifolds that admit no PL structures. See Corollary 1.8.4 and Remark 1.8.6 for such examples.

12. Rokhlin Signature Theorem. This is Theorem 1.8.1: Let M be a closed 4-dimensional PL manifold with $w_1(M) = 0 = w_2(M)$. Then the signature of M is divisible by 16.

13. Freedman's Example. This is Theorem 1.8.2: There exists a closed simply-connected topological 4-dimensional manifold V with $w_2(V) = 0$ and the signature equal to 8. This example provides the equality $\sigma_1^{TOP} = 8$.

Actually, the original Kirby–Siebenmann proof of the Main Theorem appeared before the Freedman's example and therefore did not use the last one, see Remark 1.8.6. However, as we have seen, the inequality $\sigma_1^{PL} \neq \sigma_1^{TOP}$ clarify relations between PL and topological manifolds, and thus Freedman's example should be (and is) incorporated in the exposition of the global picture.

Chapter 1

Architecture of the Proof

1.1 Some Definitions, Notation, and Conventions

1.1.1 Convention.

We agree that 0 is *not* a natural number. So, $\mathbb{N} = \{1, 2, \ldots, n, \ldots\}$.
We denote the cyclic group of order m by $\mathbb{Z}/m$.

All maps are supposed to be continuous.

All neighborhoods are supposed to be open.

All manifolds are assumed to be metrizable and separable. The word
"manifold" means "manifold without boundary" (compact or not); otherwise we say about ∂-manifold or mention the boundary explicitly.

All functional spaces are assumed to be equipped with the compact-open
topology unless the contrary is specified.

We work mainly in the class of CW-spaces. However, when we quit the
class by taking products or functional spaces, we equip the last ones with the
compactly generated topology (following Steenrod [St2] and McCord [McC],
see e.g. [Rud] for the exposition).

1.1.2 Definition. A *pointed space* is pair $(X, \{x_0\})$ where x_0 is a point
of a topological space X. We also use that notation (X, x_0) and call x_0
the *base point* of X. If we do not need to indicate the base point, we can
write $(X, *)$ (or even X if it is clear that X is a pointed space). Given two
pointed spaces (X, x_0) and (Y, y_0), a *pointed map* is a map $f : X \to Y$ such
that $f(x_0) = y_0$.

We denote the one-point space by pt.

1.1.3 Notation. Given two topological spaces X, Y, we denote by $[X, Y]$ the set of homotopy classes of maps $X \to Y$. We use the notation $[X, Y]^{\bullet}$ for the set of pointed homotopy classes of pointed maps $X \to Y$ of pointed spaces.

It is quite standard to denote by $[f]$ the homotopy class of a map f. However, frequently we do not distinguish between a map and its homotopy class, and use the same symbol, say f for a map as well as for the homotopy class. Here this does not lead to any confusion.

1.1.4 Definition. We use the term *inessential map* for a map that is homotopic to a constant map; otherwise a map is called *essential*.

We use the symbol $\simeq$ for homotopy of maps or homotopy equivalence of spaces. We use the symbol $\cong$ for bijection of sets or isomorphism of groups (rings, vector bundles, etc.). We use the symbol $:=$ for "is defined to be".

1.1.5 Notation. We use the notation π_n^S for the n-th stable homotopy group $\pi_{n+N}(S^N)$, N large.

1.1.6 Definition. We define a map f to be *proper* if $f^{-1}(C)$ is compact whenever C is compact. A map $f : X \to Y$ is *proper homotopy equivalence* if there exist a map $g : Y \to X$ and homotopies $F : gf \simeq 1_X$, $G : fg \simeq 1_Y$ such that all the four maps $f, g, F : X \times I \to X$, and $G : Y \times I \to Y$ are proper.

1.1.7 Definition. We reserve the term *bundle* for locally trivial bundles and the term *fibration* for Hurewicz fibrations.

1.1.8 Definition. Given a space F, an *F-bundle* is a bundle whose fibers are homeomorphic to F, and an *F-fibration* is a fibration whose fibers are homotopy equivalent to F.

We denote the *trivial F-bundle* $X \times F \to X$ over X by θ_X^F or merely θ^F. Also, we denote the trivial $\mathbb{R}^n$-bundle over X by θ_X^n or θ^n.

1.1.9 Remark. We do not discuss *microbundles*, because in the topological and in the PL category every n-dimensional microbundle over a good enough space X (like manifolds or locally finite finite-dimensional CW spaces) contains an $\mathbb{R}^n$-bundle over X, and these bundles are unique up to equivalence, see Kister [Kis] for the topological category and Kuiper-Lashof [KL] for the PL category. For this reason any statements on microbundles can be restated in terms of bundles. The reader should keep it in mind when we cite (quote about) something concerning microbundles.

1.1.10 Definition. Given a bundle or fibration $\xi = \{p : E \to B\}$, the space B is called the *base* of ξ and denote also by $\mathrm{bs}(\xi)$, i.e., $\mathrm{bs}(\xi) = B$. The space E is called the *total space* of ξ. Furthermore, given a space X, we set

$$\xi \times X := \{p \times 1 : E \times X \to B \times X\}.$$

1.1.11 Definition. Given two bundles $\xi = \{p : E \to B\}$ and $\eta = \{q : Y \to X\}$, a *bundle morphism* $\varphi : \xi \to \eta$ is a commutative diagram

$$
\begin{array}{ccc}
E & \xrightarrow{\ g\ } & Y \\
{\scriptstyle p}\downarrow & & \downarrow{\scriptstyle q} \\
B & \xrightarrow{\ f\ } & X.
\end{array}
$$

We say that f is the *base* of the bundle morphism φ or that φ is a *morphism over* f. We also say that g *is a map over* f. If $X = B$ and $f = 1_B$ we say that g is a map over B (and φ is a morphism over B).

Frequently we'll say just "morphism" instead of "bundle morphism" if it will not lead to confusions.

1.1.12 Definition. Given a map $f : Z \to B$ and a bundle (or fibration) $\xi = \{p : E \to B\}$, we use the notation $f^*\xi$ for the *induced bundle* over Z. Recall that the induced bundle is the bundle $f^*(\xi) := \{r : D \to Z\}$ where

$$D = \{(z, e) \in Z \times E \mid f(z) = p(e)\} \quad \text{and} \quad r(z, e) = z.$$

There is a canonical bundle morphism

$$\mathfrak{I} = \mathfrak{I}_f = \mathfrak{I}_{f,\xi} : f^*\xi \to \xi$$

given by the map

$$D \to E, \quad (z, e) \mapsto e$$

over f, see [Rud] (or [FR] where it is denoted by $\mathrm{ad}(f)$). Following [FR], we call $\mathfrak{I}_{f,\xi}$ the *adjoint morphism* of f, or just the *f-adjoint* morphism. Furthermore, given a bundle morphism $\varphi : \xi \to \eta$ with the base f, there exists a unique bundle morphism

$$c(\varphi) : \xi \to f^*\eta$$

over the base of ξ such that the composition

$$\xi \xrightarrow{\;c(\varphi)\;} f^*\eta \xrightarrow{\;\mathfrak{I}_{f,\eta}\;} \eta$$

coincides with φ. Following [FR], we say that $c(\varphi)$ the *correcting morphism* for φ.

Given a subspace A of a space X and a bundle ξ over X, we denote by $\xi|_A$ the bundle $i^*\xi$ where $i : A \subset X$ is the inclusion.

1.1.13 Definition. Given a map $p : E \to B$ and a map $f : X \to B$, a *p-lifting of f* is any map $g : X \to E$ with $pg = f$. Two p-liftings g_0, g_1 of f are *vertically homotopic* if there exists a homotopy $G : X \times I \to E$ between g_0 and g_1 such that $pg_t = f$ for all $t \in I$. The set of vertically homotopic p-liftings of f is denoted by $[\mathrm{Lift}_p\, f]$.

1.1.14 Notation. We denote by p_k, w_k and L_k the Pontryagin, Stiefel–Whitney, and Hirzebruch characteristic classes, respectively. We denote by $\sigma(M)$ the signature of a manifold M. See [MS, Hirz] for the definitions.

1.2 Principal Fibrations

Recall that an H-space is a space F with a base point f_0 and a multiplication map $\mu : F \times F \to F$ such that f_0 is the strict unit, i.e., $f = \mu(f, f_0)$ and $f = \mu(f_0, f)$ for all $f \in F$. For details, see [BV].

1.2.1 Definition. (a) Let (F, f_0) be an H-space with the multiplication $\mu : F \times F \to F$. A *principal F-fibration* is an F-fibration $p : E \to B$ equipped with a map $m : E \times F \to E$ (the F-action on E) such that the following holds:

(i) the diagrams

$$
\begin{array}{ccc}
E \times F \times F \xrightarrow{\;m \times 1\;} E \times F & \qquad & E \times F \xrightarrow{\;m\;} E \\
\quad\downarrow{\scriptstyle 1 \times \mu} \qquad\qquad \downarrow{\scriptstyle m} & & \quad\downarrow{\scriptstyle p_1} \qquad\qquad \downarrow{\scriptstyle p} \\
E \times F \xrightarrow{\;\;m\;\;} E & & E \xrightarrow{\;\;p\;\;} B
\end{array}
$$

commute;

(ii) for every $e_0 \in E$, the map

$$F \longrightarrow p^{-1}(p(e_0)), \quad f \mapsto m(e_0, f)$$

is a homotopy equivalence.

(b) A *trivial* principal F-fibration is the fibration $p_2 : X \times F \to F$ with an F-action $m : E \times F \to E$ of the form

$$m : X \times F \times F \to X \times F, \quad m(x, f_1, f_2) = (x, \mu(f_1, f_2)).$$

It is easy to see that if the fibration η is induced from a principal fibration ξ then η turns into a principal fibration in a canonical way.

1.2.2 Definition. Let $\pi_1 : E_1 \to B$ and $\pi_2 : E_2 \to B$ be two principal F-fibrations over the same base B. We say that a map $h : E_1 \to E_2$ is an *F-equivariant map over B* if h is a map over B and the diagram

$$
\begin{array}{ccc}
E_1 \times F & \xrightarrow{\ h \times 1\ } & E_2 \times F \\
{\scriptstyle m_1} \downarrow & & \downarrow {\scriptstyle m_2} \\
E_1 & \xrightarrow{\ h\ } & E_2
\end{array}
$$

commutes up to homotopy over B.

Note that, for every $b \in B$, the map

$$h_b : \pi_1^{-1}(b) \to \pi_2^{-1}(b), \quad h_b(x) = h(x)$$

is a homotopy equivalence.

Now, let $p : E \to B$ be a principal F-fibration, and let $f : X \to B$ be an arbitrary map. Given a p-lifting $g : X \to E$ of f and a map $u : X \to F$, consider the map

$$g_u : X \xrightarrow{\ \Delta\ } X \times X \xrightarrow{\ g \times u\ } E \times F \xrightarrow{\ m\ } E.$$

It is easy to see that the correspondence $(g, u) \mapsto g_u$ yields a well-defined map (right action)

$$[\mathrm{Lift}_p\, f] \times [X, F] \to [\mathrm{Lift}_p\, f]. \tag{1.2.1}$$

In particular, for every p-lifting g of f the correspondence $u \mapsto g_u$ induces a map

$$T_g : [X, F] \to [\mathrm{Lift}_p\, f].$$

1.2.3 Proposition. *If F is a homotopy associative H-space with a homotopy inversion, then $[X, F]$ is a group.*

Proof. See e.g. [Wh1, Chapter III, §4], cf. also [FFG, Sw].   □

1.2.4 Theorem. *Let $\xi = \{p : E \to B\}$ be a principal F-fibration, and let $f : X \to B$ be a map where X is assumed to be paracompact and locally contractible. Assume that F is a homotopy associative H-space with a homotopy inversion. If $[\mathrm{Lift}_p f] \neq \emptyset$ then the above action (1.2.1) of the group $[X, F]$ is free and transitive. In particular, for every p-lifting $g : X \to E$ of f the map T_g is a bijection.*

Proof. We start with the following lemma.

1.2.5 Lemma. *The theorem holds if $X = B$, $f = 1_X$ and ξ is the trivial principal F-fibration.*

Proof. In this case every p-lifting $g : X \to X \times F$ of $f = 1_X$ is completely determined by the map

$$\overline{g} : X \xrightarrow{\ g\ } X \times F \xrightarrow{\ p_2\ } F.$$

In other words, we have the bijection $[\mathrm{Lift}_p f] \cong [X, F]$, and under this bijection the action (1.2.1) turns into the group multiplication

$$[X, F] \times [X, F] \to [X, F].$$

Now the result follows since $[X, F]$ is a group. The lemma is proved.

We complete the proof of the theorem. Consider the induced fibration $f^*\xi = \{q : Y \to X\}$ and note that there is an $[X, F]$-equivariant bijection

$$[\mathrm{Lift}_p f] \cong [\mathrm{Lift}_q 1_X]. \tag{1.2.2}$$

Now, suppose that $[\mathrm{Lift}_p f] \neq \emptyset$ and take a p-lifting g of f. We regard Y as the subset of $X \times E$, and define the F-equivariant map

$$h : X \times F \to Y, \quad h(x, a) = (x, g(x)a), \ x \in X, a \in F.$$

It is easy to see that the diagram

$$
\begin{array}{ccc}
X \times F & \xrightarrow{\ h\ } & Y \\
{\scriptstyle p_1}\big\downarrow & & \big\downarrow{\scriptstyle q} \\
X & =\!=\!=\!= & X
\end{array}
$$

commutes, i.e., h is a map over X. Since X is a locally contractible para-compact space, and because of a theorem of Dold [Dold], there exists a map $k : Y \to X \times F$ over X which is homotopy inverse over X to h. It is easy to see that k is an equivariant map over X. Indeed, if $m_1 : X \times F \times F \to X \times F$ and $m_2 : Y \times F \to Y$ are the corresponding F-actions then

$$m_1(k \times 1) \simeq khm_1(k \times 1) \simeq km_2(h \times 1)(k \times 1) \simeq km_2(hk \times 1) \simeq km_2,$$

where $\simeq$ denotes the homotopy over X.

In particular, there is an $[X, F]$-equivariant bijection

$$[\mathrm{Lift}_q \, 1_X] \cong [\mathrm{Lift}_{p_1} \, 1_X]$$

where $p_1 : X \times F \to X$ is the projection. Now we compose this bijection with (1.2.2) and get $[X, F]$-equivariant bijections

$$[\mathrm{Lift}_p \, f]] \cong [\mathrm{Lift}_q \, 1_X] \cong [\mathrm{Lift}_{p_1} \, 1_X],$$

and the result follows from Lemma 1.2.5. $\qquad\square$

1.2.6 Example. If $p : E \to B$ is an F-fibration then $\Omega p : \Omega E \to \Omega B$ is a principal ΩF-fibration. Here Ω denotes the loop functor.

1.3 Preliminaries on Classifying Spaces

Here we give a brief recollection on $\mathbb{R}^n$-bundles, spherical fibrations, and their classifying spaces. For more detailed survey, see [Rud, Chapter IV].

1.3.1 Definition. We define a *topological $\mathbb{R}^n$-bundle* over a space B to be an $\mathbb{R}^n$-bundle $\xi = \{p : E \to B\}$ equipped with a fixed section $s : B \to E$ (the zero section) assumed to be a cofibration. Given two topological $\mathbb{R}^n$-bundles $\xi = \{p : E \to B\}$ and $\eta = \{q : Y \to X\}$, we define a *topological $\mathbb{R}^n$-morphism* $\varphi : \xi \to \eta$ to be a commutative diagram (a bundle morphism)

$$
\begin{array}{ccc}
E & \xrightarrow{\ g\ } & Y \\
{\scriptstyle p}\downarrow & & \downarrow{\scriptstyle q} \\
B & \xrightarrow{\ f\ } & X
\end{array}
\qquad (1.3.1)
$$

where φ respects the sections $(gs_\xi = s_\eta f)$ and induces a homeomorphism on each of the fibers. The last one means that, for every $b \in B$, the map

$$g_b : \mathbb{R}^n = p^{-1}(b) \to q^{-1}(f(b)) = \mathbb{R}^n,$$
$$g_b(a) = g(a) \text{ for all } a \in p^{-1}(b)$$

is a homeomorphism.

A topological $\mathbb{R}^n$-morphism is a topological $\mathbb{R}^n$-*isomorphism* if the above-mentioned g is a homeomorphism.

We define two topological $\mathbb{R}^n$-morphisms $\varphi_0, \varphi_1 : \xi \to \eta$ to be *bundle homotopic* if there exists a topological $\mathbb{R}^n$-morphism $\Phi : \xi \times I \to \eta$ such that $\Phi|_\xi \times \{i\} = \varphi_i, i = 0, 1$.

A topological $\mathbb{R}^n$-morphism $\varphi : \xi \to \eta$ is a *bundle homotopy equivalence* if there exists a topological $\mathbb{R}^n$-morphism $\psi : \eta \to \xi$ such that $\varphi\psi$ and $\psi\varphi$ are bundle homotopic to the corresponding identity maps.

Frequently, we will just say "homotopy" instead of "bundle homotopy", etc. if this does not lead to confusions.

1.3.2 Theorem–Definition. *There exists a topological $\mathbb{R}^n$-bundle γ^n_{TOP} with the following* universal property: *For every topological $\mathbb{R}^n$-bundle ξ over a CW-space B, every CW-subspace A of B and every morphism*

$$\psi : \xi_A \to \gamma^n_{TOP}$$

of topological $\mathbb{R}^n$-bundles, there exists a morphism $\varphi : \xi \to \gamma^n_{TOP}$ which is an extension of ψ. The base of γ^n_{TOP} is called the classifying space *for topological $\mathbb{R}^n$-bundles and denoted by $BTOP_n$.*

We can regard topological $\mathbb{R}^n$-bundles as $(TOP_n, \mathbb{R}^n)$-bundles, i.e., $\mathbb{R}^n$-bundles with the structure group TOP_n. Here TOP_n is the topological group of self-homeomorphism $f : \mathbb{R}^n \to \mathbb{R}^n, f(0) = 0$ equipped with the compact-open topology. The classifying space $BTOP_n$ of the group TOP_n turns out to be a classifying space for topological $\mathbb{R}^n$-bundles.

Consider a topological $\mathbb{R}^n$-bundle ξ over a CW space B. By the definition of universal bundle, there exists a topological $\mathbb{R}^n$-morphism $\varphi : \xi \to \gamma^n_{TOP}$. We call such φ a *classifying morphism* for ξ. The base $f : B \to BTOP_n$ of φ is called a *classifying map* for ξ. It is clear that ξ is isomorphic over B to $f^*\gamma^n_{TOP}$.

1.3.3 Proposition. *Any two classifying morphisms*

$$\varphi_0, \varphi_1 : \xi \to \gamma_{TOP}^n$$

for ξ are homotopic. In particular, a classifying map f for ξ is determined by ξ uniquely up to homotopy.

Proof. Put $A = (\mathrm{bs}\,\xi) \times \{0,1\}$ and apply the universal property 1.3.2 to $\xi \times I$. $\qquad\qquad\qquad\qquad\qquad\qquad\qquad\qquad\qquad\qquad\qquad\qquad\qquad\Box$

It follows from the previous proposition that the space $BTOP_n$ is defined uniquely up to weak homotopy equivalence.

1.3.4 Remark. This is important to understand the difference between classifying maps and classifying morphisms. Every morphism $\xi \to \gamma_{TOP}^n$ is a classifying morphism, while not every map $X \to BTOP_n$ is a classifying map (consider the case of the trivial bundle over X).

1.3.5 Definition. A *PL $\mathbb{R}^n$-bundle* is a topological $\mathbb{R}^n$-bundle $\xi = \{p : E \to B\}$ such that E and B are polyhedra, and $p : E \to B$ and $s : B \to E$ PL maps. Furthermore, we require that, for every simplex $\Delta \subset B$, there is a PL homeomorphism $h : p^{-1}(\Delta) \cong \Delta \times \mathbb{R}^n$ with $h(s(\Delta)) = \Delta \times \{0\}$.

A *PL morphism* of PL $\mathbb{R}^n$-bundles is a topological $\mathbb{R}^n$-morphism where the maps g and f in (1.3.1) are PL maps.

1.3.6 Theorem–Definition. *There exists a universal PL $\mathbb{R}^n$-bundle γ_{PL}^n. This means that the* universal property 1.3.2 *remains valid if we replace γ_{TOP}^n by γ_{PL}^n and "topological $\mathbb{R}^n$ bundle" by "PL $\mathbb{R}^n$-bundle" there. The base BPL_n of γ_{PL}^n is called a* classifying space *for PL $\mathbb{R}^n$-bundles, and it is defined uniquely up to weak homotopy equivalence.*

Note that BPL_n can also be regarded as the classifying space of a certain group PL_n (which is constructed as the geometric realization of a certain simplicial group), [KL, LR1].

This is worthy to mention that BPL_n can be chosen to be a locally finite simplicial complex, [KS2, Essay IV, §8].

1.3.7 Definition. A *homotopy S^{n-1}-fibration* is defined to be a fibration $p : E \to B$ whose fibers are homotopy equivalent to S^{n-1}. In future, frequently we use shorter term "S^{n-1}-fibration" or just "spherical fibration" if there is no danger of confusion.

Given two homotopy S^{n-1}-fibrations $\xi = \{p : E \to B\}$ and $\eta = \{q : Y \to X\}$, we define a morphism of spherical fibrations $\varphi : \xi \to \eta$ to be a commutative diagram

$$
\begin{array}{ccc}
E & \xrightarrow{\;g\;} & Y \\
{\scriptstyle p}\downarrow & & \downarrow{\scriptstyle q} \\
B & \xrightarrow{\;f\;} & X
\end{array}
\qquad (1.3.2)
$$

where for each $b \in B$ the map

$$
g_b : p^{-1}(b) \to q^{-1}(f(b))
$$

should be a homotopy equivalence.

1.3.8 Theorem–Definition. *There exists a universal homotopy S^{n-1}-fibration γ_G^n. This means that the universal property 1.3.2 remains valid if we replace γ_{TOP}^n by γ_G^n and "topological $\mathbb{R}^n$ bundle" by homotopy S^n-fibration" in 1.3.2. The base BG_n of γ_G^n is called the* classifying space *for homotopy S^{n-1}-fibrations, and it is defined uniquely up to weak homotopy equivalence.*

1.3.9 Remark. The space BG_n can also be regarded as the classifying space for the monoid G_n of homotopy self-equivalences $S^{n-1} \to S^{n-1}$ equipped with compact-open topology.

1.3.10 Notation. We need also to recall the space BO_n which classifies n-dimensional vector bundles. This is a well-known space, see e.g., [MS]. The universal vector bundle over BO_n is denoted by γ_O^n.

1.3.11 Notation. We regard γ_{PL}^n as the (underlying) topological $\mathbb{R}^n$-bundle and get the classifying morphism

$$
\omega = \omega_{TOP}^{PL}(n) : \gamma_{PL}^n \to \gamma_{TOP}^n.
$$

The base of the morphism is denoted by

$$
\alpha = \alpha_{TOP}^{PL}(n) : BPL_n \to BTOP_n.
$$

1.3.12 Construction–Definition. Any n-dimensional vector bundle over a polyhedron X has a canonical structure on PL $\mathbb{R}^n$-bundle over X. So, we can regard γ_O^n as PL bundle and get a (forgetful) classifying morphism

$$
\omega_{PL}^O(n) : \gamma_O^n \to \gamma_{PL}^n.
$$

We denote by $\alpha_{PL}^{O}(n) : BO_n \to BPL_n$ the base of $\omega_{PL}^{O}(n)$.

We can define the morphism $\omega_{PL}^{O}(n)$ and a map $\alpha_{PL}^{O}(n)$ in a similar way.

The case of BG is more delicate. Any topological $\mathbb{R}^n$-bundle can be converted to homotopy S^{n-1}-fibration via deletion of zero section from B. In detail, we take a topological $\mathbb{R}^n$-bundle $\xi = \{p : E \to B\}$ and obtain the S^{n-1}-fibration denoted by

$$\xi^{\nabla} := \{p' : E \setminus s(B) \to B\}.$$

In particular, we have a homotopy S^{n-1}-fibration $(\gamma_{TOP}^n)^{\nabla}$ over $BTOP_n$. So, there is a classifying morphism

$$\omega_G^{TOP}(n) : \gamma_{TOP}^n \to \gamma_G^n.$$

We denote by $\alpha_G^{TOP}(n) : BTOP_n \to BG_n$ the base of $\omega_G^{TOP}(n)$.

Thus, we have a sequence of forgetful maps

$$BO_n \xrightarrow{\alpha'} BPL_n \xrightarrow{\alpha''} BTOP_n \xrightarrow{\alpha'''} BG_n \qquad (1.3.3)$$

where $\alpha' = \alpha_{PL}^{O}(n)$, etc.

1.3.13 Definition. (a) Given an F-bundle $\xi = \{p : E \to B\}$ and an F'-bundle $\xi' = \{p' : E' \to B'\}$, we define the product $\xi \times \xi'$ to be the $F \times F'$-bundle

$$p \times p' : E \times E' \to B \times B'.$$

In particular, given an $\mathbb{R}^m$-bundle ξ and an $\mathbb{R}^n$-bundle η we have the $\mathbb{R}^{m+n}$-bundle $\xi \times \eta$.

(b) Given an $\mathbb{R}^m$-bundle ξ and an $\mathbb{R}^n$-bundle η over the same space X, the *Whitney sum* of ξ and η is the $\mathbb{R}^{m+n}$-bundle

$$\xi \oplus \eta := d^*(\xi \times \eta)$$

where $d : X \to X \times X$, $d(x) = (x, x)$ is the diagonal map.

(c) We denote the *fiberwise join*, or *bundle join* (see [Rud, IV.1.4], [FR]) of homotopy spherical fibrations σ and τ by $\sigma * \tau$. Given an $\mathbb{R}^m$-bundle ξ and an $\mathbb{R}^n$-bundle η, note that $\xi^{\nabla} * \eta^{\nabla}$ is fiber homotopy equivalent to $(\xi \times \eta)^{\nabla}$.

We define the analog of Whitney sum for spherical fibrations. Namely, if σ and τ are two spherical fibration of the same base X we define the spherical fibration $\sigma \dagger \tau := d^*(\sigma * \tau)$. In particular, given an $\mathbb{R}^m$-bundle ξ and an $\mathbb{R}^n$-bundle η over the same base, we have $(\xi \oplus \eta)^\nabla$ is fiber homotopy equivalent to $\xi^\nabla \dagger \eta^\nabla$.

1.3.14 Definition. We define

$$r_n = r_n^{TOP} : BTOP_n \to BTOP_{n+1}$$

to be the map which classifies topological $\mathbb{R}^{n+1}$-bundle

$$\gamma_{TOP}^n \oplus \theta_{BTOP_n}^1.$$

The maps $r_n^G : BG_n \to BG_{n+1}$, $r_n^{PL} : BPL_n \to BPL_{n+1}$, and $r_n^O : BO_n \to BO_{n+1}$ are defined in a similar way.

We can also regard the map $r_n : BTOP_n \to BTOP_{n+1}$ as a map induced by the standard inclusion $TOP_n \subset TOP_{n+1}$. Similarly, we can regard $r_n^G : BG_n \to BG_{n+1}$ as the map induced by the standard inclusion $G_n \subset G_{n+1}$, see [MM, p. 45].

Now we consider the sequences

$$BO_n \xrightarrow{\ r_n\ } BPL_n \longrightarrow BTOP_n \longrightarrow BG_n$$

as $n \to \infty$. Remember that the classifying spaces BG, etc. are defined up to weak homotopy equivalence. In order to pass to the limit, we should take an accurate choice of the classifying spaces in the corresponding homotopy classes. In detail, we do the following.

1.3.15 Definition and Notation. Choose classifying spaces $B'G_n, n \in \mathbb{N}$ for S^{n-1}-fibrations (i.e., in the weak homotopy type BG_n) and consider the maps $r_n^G : B'G_n \to B'G_{n+1}$ as above. We can assume that every $B'G_n$ is a CW-complex and every r_n is a cellular map. We define BG to be the telescope (homotopy direct limit) of the sequence

$$\cdots \longrightarrow B'G_n \xrightarrow{\ r_n\ } B'G_{n+1} \longrightarrow \cdots,$$

see e.g., [Rud, Definition I.3.19]. Furthermore, from now on and forever we reserve the notation BG_n for the classifying space that is the telescope of the finite sequence

$$\cdots \longrightarrow B'G_{n-1} \xrightarrow{\ r_{n-1}\ } B'G_n.$$

So, we have the sequence (filtration)

$$\cdots \subset BG_n \subset BG_{n+1} \subset \cdots .$$

Hence, $BG = \bigcup BG_n$ and BG_n is closed in BG. Moreover, BG has the direct limit topology with respect to the filtration $\{BG_n\}$. Furthermore, if $f : K \to BG$ is a map of a compact space K then there exists n such that $f(K) \subset BG_n$.

1.3.16 Definition and Notation. For every n consider a CW-space $B'TOP_n$ in the weak homotopy type $BTOP_n$ and define $B''TOP$ to be the telescope of the sequence

$$\cdots \longrightarrow B'TOP_n \xrightarrow{r_n} B'TOP_{n+1} \longrightarrow \cdots .$$

Furthermore, we define $B''TOP_n$ to be the telescope of the finite sequence

$$\cdots \longrightarrow B'TOP_{n-1} \xrightarrow{r_{n-1}} B'TOP_n.$$

So, we have the diagram

$$
\begin{array}{ccccccc}
\cdots \subset B''TOP_n & \subset & B''TOP_{n+1} & \subset \cdots \subset & B''TOP \\
\downarrow & & \downarrow & & \downarrow{\scriptstyle p} & & (1.3.4)\\
\cdots \subset \quad BG_n & \subset & BG_{n+1} & \subset \cdots \subset & BG
\end{array}
$$

where the map p is induced by maps $\alpha_F^{TOP}(n)$. Now we apply the Serre construction and replace every vertical map in the diagram (1.3.4) by its fibrational substitute. Namely, we set

$$BTOP = \{(x,\omega) \mid x \in B''TOP,\, \omega \in (BG)^I,\, p(x) = \omega(0)\}$$

and define $\alpha_F^{TOP} : BTOP \to BG$ by setting $\alpha_F^{TOP}(x,\omega) = \omega(1)$. Finally, we set

$$BTOP_n = \{(x,\omega) \in BTOP \mid x \in B''TOP_n,\, \omega \in (BG_n)^I\}$$

and get the commutative diagram

$$
\begin{array}{ccccccc}
\cdots \subset BTOP_n & \subset & BTOP_{n+1} & \subset \cdots \subset & BTOP \\
\downarrow & & \downarrow & & \downarrow{\scriptstyle p} \\
\cdots \subset \quad BG_n & \subset & BG_{n+1} & \subset \cdots \subset & BG
\end{array}
$$

where all the vertical maps are fibrations.

Now it is clear how to proceed and get the diagram

$$
\begin{array}{ccccccc}
\cdots\subset & BO_n & \subset & BO_{n+1} & \subset\cdots\subset & BO \\
& \downarrow & & \downarrow & & \downarrow{\scriptstyle\alpha_{PL}^{O}} \\
\cdots\subset & BPL_n & \subset & BPL_{n+1} & \subset\cdots\subset & BPL \\
& \downarrow & & \downarrow & & \downarrow{\scriptstyle\alpha_{TOP}^{PL}} \\
\cdots\subset & BTOP_n & \subset & BTOP_{n+1} & \subset\cdots\subset & BTOP \\
{\scriptstyle\alpha_F^{TOP}(n)}\downarrow & & \downarrow & & & \downarrow{\scriptstyle\alpha_F^{TOP}} \\
\cdots\subset & BG_n & \subset & BG_{n+1} & \subset\cdots\subset & BG
\end{array}
\qquad (1.3.5)
$$

where all the vertical maps are fibrations. Moreover, each of limit spaces has the direct limit topology with respect to the corresponding filtration, and every compact subspace of, say, BO is contained in some BO_n.

1.3.17 Convention. Let ξ classify a map $f_n : X \to BG_n$ (or BO_n, etc.). It is convenient for us to speak about $n = \infty$ and write that a map $f : X \to BG$ classify ξ if f can be expressed as

$$ f : X \xrightarrow{\ f_n\ } BG_n \xrightarrow{\ \subset\ } BG. $$

1.3.18 Notation. Take a point $b \in BTOP$, put

$$ (TOP/PL)_b := (\alpha_{TOP}^{PL})^{-1}(b) $$

to be the fiber of $\alpha = \alpha_{TOP}^{PL}$, and put

$$ \beta = \beta_b : (TOP/PL)_b \to BPL $$

to be the inclusion of the fiber. In future we allow us to omit the subscript b and speak about the fibration

$$ TOP/PL \xrightarrow{\ \beta_{TOP}^{PL}\ } BPL \xrightarrow{\ \alpha_{TOP}^{PL}\ } BTOP. \qquad (1.3.6) $$

This will not lead to confusions because, if we choose another point $b' \in BTOP$, then the maps β_b and $\beta_{b'}$ occur to be homotopy equivalent. We also use the notation TOP/PL for the homotopy fiber of the map $\alpha : BPL \to BTOP$.

The homotopy fiber of $\alpha_{PL}^O : BO \to BPL$ is denoted by PL/O, the fiber of α_{TOP}^F is denoted by G/TOP, etc. Similarly, the homotopy fiber of the composition, say,

$$ \alpha_F^{PL} := \alpha_F^{TOP} \circ \alpha_{TOP}^{PL} : BPL \to BG $$

is denoted by G/PL. In particular, we have a fibration

$$TOP/PL \xrightarrow{\ a\ } G/PL \xrightarrow{\ b\ } G/TOP. \qquad (1.3.7)$$

Finally, note that $G/TOP = \bigcup G_n/TOP_n$ where G_n/TOP_n denotes the fiber of the fibration $BTOP_n \to BG_n$, and G/TOP has the direct limit topology with respect to the filtration $\{G_n/TOP_n\}$. The same holds for other "homogeneous spaces" $G/PL, TOP/PL$, etc.

Because of the well-known results of Milnor [Mi1], all these "homogeneous spaces" have the homotopy type of CW-spaces. Furthermore, all the spaces BPL, $BTOP$, G/PL, TOP/PL, etc. are infinite loop spaces, and the maps like in (1.3.6) and (1.3.7) are infinite loop maps, see [BV]. In particular, the classifying spaces BPL, etc. are homotopy associative and invertible H-spaces, and the fibrations (1.3.6), (1.3.7), etc. are principal fibrations.

We mention also the following useful fact.

1.3.19 Lemma. *Let Z denote one of the symbols O, PL, G. The above described map $BZ_n \to BZ_{n+1}$ induces an isomorphism of homotopy groups in dimensions $\leqslant n - 1$ and an epimorphism in dimension n.*

Proof. For $Z = O$ and $Z = G$ it is well known, see e.g., [Br2], for $Z = PL$ it can be found in [HW]. □

1.3.20 Remark. An analog of Lemma 1.3.19 holds for TOP as well, see Remark 1.4.13.

1.3.21 Remark. Let F_n denote the topological monoid of pointed self-equivalences $S^n \to S^n$. Then the classifying space BF_n of F_n classifies sectioned S^n-fibrations. The obvious forgetful map $TOP_n \to G_n$ (regard self-homeomorphism as a self-equivalence) induces a map

$$BTOP_n \longrightarrow BF_n$$

of classifying spaces. In the language of bundles, this map converts a topological $\mathbb{R}^n$-bundle into a sectioned spherical fibration.

We can also consider the space BF by tending n to ∞. In particular, we have the spaces F/PL and F/TOP.

There is an obvious forgetful map $F_n \to G_{n+1}$ (ignore section), and it turns out that the induced map $BF \to BG$ (as $n \to \infty$) is a homotopy equivalence, see e.g., [MM, Chapter 3]. In particular,

$$G/PL \simeq F/PL. \qquad (1.3.8)$$

1.4 Structures on Manifolds and Bundles

1.4.1 Definition. A *PL atlas* on a topological manifold is an atlas such that all the transition maps are PL ones. We define a PL manifold as a topological manifold with a maximal PL atlas. Furthermore, given two PL manifolds M and N, we say that a homeomorphism $h : M \to N$ a *PL homeomorphism* if h is a PL map. (One can prove that in this case h^{-1} is a PL map as well, [Hud].)

1.4.2 Definition. (a) We define a ∂_{PL}-*manifold* to be a topological manifold whose boundary ∂M is a PL manifold. In particular, every closed topological manifold is a ∂_{PL}-manifold. Furthermore, every PL manifold with boundary can be regarded as a ∂_{PL}-manifold.

(b) Let M be a ∂_{PL}-manifold. A *PL structure on M* is a homeomorphism $h : V \to M$ such that V is a PL manifold and h induces a PL homeomorphism $\partial V \to \partial M$ of boundaries (or, equivalently, PL homeomorphism of the corresponding collars). Two PL structures $h_i : V_i \to M, i = 0, 1$ are *concordant* if there exist a PL manifold W and a homeomorphism $H : W \to M \times I$ (a concordance) such that

$$H^{-1}(M \times \{i\}) = V_i, \ H|_{V_i} = h_i : V_i \to M, \quad i = 0, 1$$

and, moreover, the restricted homeomorphism

$$H : H^{-1}(\partial M \times I) \to \partial M \times I$$

coincides with a PL homeomorphism

$$H|_{\partial V_0} \times 1 : \partial V_0 \times I \to \partial M \times I.$$

Clearly, "to be concordant" is an equivalence relation, called also concordance. We denote by $\mathcal{T}_{PL}(M)$ the set of all concordance classes of PL structures on M.

(c) If M on its own is a PL manifold then $\mathcal{T}_{PL}(M)$ contains a distinguished element: the concordance class of 1_M. We call it the *trivial element of $\mathcal{T}_{PL}(M)$*.

1.4.3 Remarks. 1. Clearly, every PL structure on M equips M with a certain PL atlas. Conversely, if we equip M with a certain PL atlas then the identity map can be regarded as a PL structure on M.

2. (Concordance and Isotopy.) Recall that two homeomorphisms $h_0, h_1 : X \to Y$ are *isotopic* if there exists a homeomorphism $J : X \times I \to Y \times I$ (isotopy) such that $p_2 J : X \times I \to Y \times I \to I$ coincides with $p_2 : X \times I \to I$.

Given $A \subset X$, we say that h_0 and h_1 are isotopic rel A if there exists an isotopy H such that $J(a,t) = (h_0(a), t)$ for every $a \in A$ and every $t \in I$. In particular, if two PL structures $h_0, h_1 : V \to M$ are isotopic rel ∂V then they are concordant.

3. (Concordance and Mapping Cylinders of Isotopies.) Consider two PL structures $h_i : V_i \to M, i = 0, 1$ and let $\varphi : V_0 \to V_1$ be a PL homeomorphism such that $h_1\varphi$ and h_0 are isotopic via an isotopy $J : V_0 \times I \to M \times I$. Let W denote the mapping cylinder of φ,

$$W = (V_0 \times [0,1] \cup V_1/) \sim, \text{where } (v,1) \sim h_1\varphi(v), \ v \in V_0.$$

Then W is a PL manifold (because φ is a PL homeomorphism) and produces the homeomorphism

$$W \to M \times I, \quad (v,t) \mapsto (J(v,t), t)$$

that is a concordance between h_0 and h_1.

This is remarkable that there is a partial converse to 1.4.3(2).

1.4.4 Theorem. *Let $M, \dim M \geqslant 5$ be a PL manifold. Suppose that there is a concordance between $h_0 : V_0 \to M$ and $h_1 : V_1 \to M$, as in Definition 1.4.2(b). Then there exists a PL homeomorphism $\tau : V_0 \to V_1$ such that $h_1\tau$ and h_0 are isotopic. In particular, V_0 and V_1 are PL homeomorphic.*

Proof. See [Q4, p. 305], cf. also [KS2, Essay I, §4]. $\square$

1.4.5 Remark. There are examples of two non-concodant PL structures $h_i : V_i \to M, i = 0, 1$ such that V_0 and V_1 are PL homeomorphic. We are not able to give such examples here, but we do it later, see Remark 1.4.14(2) and Example 3.5.3.

1.4.6 Definition (cf. [Br2, Rud]). Given a topological $\mathbb{R}^n$-bundle ξ, define a *PL prestructure* on ξ to be a topological $\mathbb{R}^n$-morphism $\varphi : \xi \to \gamma_{PL}^n$. We define a *PL structure on* ξ to be a bundle homotopy class of PL prestructures of ξ.

Let $f_n : X \to BTOP_n$ classify a topological $\mathbb{R}^n$-bundle ξ, and assume that there is an $\alpha_{TOP}^{PL}(n)$-lifting

$$g_n : X \to BPL_n$$

of f_n. Take the g_n-adjoint classifying morphism

$$\mathfrak{J} = \mathfrak{J}_{g_n} : g_n^* \gamma^n \to \gamma^n$$

and consider the morphism

$$\xi \cong f_n^* \gamma_{TOP}^n = g_n^* \alpha(n)^* \gamma_{TOP}^n = g_n^* \gamma_{PL}^n \xrightarrow{\jmath} \gamma_{PL}^n$$

where $\alpha(n) := \alpha_{TOP}^{PL}(n)$. This morphism $\xi \to \gamma_{PL}^n$ is a PL prestructure on ξ. It is easy to see that we have a correspondence

$$[\mathrm{Lift}_{\alpha(n)} \, f_n] \longrightarrow \{\mathrm{PL \ structures \ on} \ \xi\}. \qquad (1.4.1)$$

1.4.7 Theorem. *The correspondence* (1.4.1) *is a bijection.*

Proof. See [Rud, Theorem IV.2.3], cf. also [Br2, Chapter II, §4]. □

  Consider now the map

$$f : X \xrightarrow{\quad f_n \quad} BTOP_n \subset BTOP$$

and the map $\alpha = \alpha_{TOP}^{PL} : BPL \to BTOP$ as in (1.3.5). Every $\alpha(n)$-lifting $g_n : X \to BPL_n$ of f_n gives us the α-lifting

$$X \xrightarrow{\quad g_n \quad} BPL_n \longrightarrow BPL$$

of f. So, we have a correspondence

$$u_\xi : \{\mathrm{PL \ structures \ on} \ \xi\} \longrightarrow [\mathrm{Lift}_{\alpha(n)} \, f_n] \longrightarrow [\mathrm{Lift}_\alpha \, f] \qquad (1.4.2)$$

where the first map is the inverse to (1.4.1). Furthermore, there is a canonical map

$$v_\xi : \{\mathrm{PL \ structures \ on} \ \xi\} \longrightarrow \{\mathrm{PL \ structures \ on} \ \xi \oplus \theta^1\},$$

and these maps respect the maps u_ξ, i.e., $u_{\xi \oplus \theta^1} v_\xi = u_\xi$. So, we have the map

$$\lim_{n \to \infty} \{\mathrm{PL \ structures \ on} \ \xi \oplus \theta^n\} \longrightarrow [\mathrm{Lift}_\alpha \, f] \qquad (1.4.3)$$

where lim means the direct limit of the sequence of sets.

1.4.8 Proposition. *If X is a finite CW-space then the map* (1.4.3) *is a bijection.*

Proof. The surjectivity follows since every compact subset of $BTOP$ is contained in some $BTOP_n$. Similarly, every map $X \times I \to BPL$ passes through some BPL_n, and therefore the injectivity holds. □

  The space TOP/PL is a homotopy associative and homotopy invertible H-space, and hence the set $[X, TOP/PL]$ has a natural group structure. Here the neutral element is the homotopy class of inessential map

$X \to TOP/PL$. Now, consider a principal F-fibration $F \to E \to B$ as in Definition 1.2.1 and apply it to the case

$$TOP/PL \xrightarrow{\ \beta\ } BPL \xrightarrow{\ \alpha\ } BTOP.$$

Then for every map $f : X \to BTOP$ we have a right action

$$r : [\mathrm{Lift}_\alpha\, f] \times [X, TOP/PL] \longrightarrow [\mathrm{Lift}_\alpha\, f].$$

1.4.9 Proposition. *Suppose that the map $f : X \to BTOP$ lifts to BPL. Then the action r is transitive. Furthermore, for every α-lifting g of f the map*

$$[X, TOP/PL] \longrightarrow [\mathrm{Lift}_\alpha\, f], \quad \varphi \mapsto r(g, \varphi)$$

is a bijection.

Proof. See Theorem 1.2.4.   □

Note that, in view of Propositions 1.4.8 and 1.4.9, if a topological bundle ξ admits a PL structure then the bijection (1.4.3) turns into the bijection

$$\lim_{n \to \infty} \{\text{PL structures on } \xi \oplus \theta^n\} \longrightarrow [X, TOP/PL] \qquad (1.4.4)$$

provided that X is a finite CW space.

1.4.10 Definition. Let M be a ∂_{PL}-manifold. A *homotopy triangulation* of M is a homotopy equivalence $h : (V, \partial V) \to (M, \partial M)$ such that V is a PL manifold and $h|_{\partial V} : \partial V \to \partial M$ is a PL homeomorphism. Two homotopy triangulations $h_i : V_i \to M, i = 0, 1$ are *equivalent* if there exists a PL homeomorphism $\varphi : V_0 \to V_1$ such that $h_1\varphi$ is homotopic to h_0 rel ∂V_0. In detail, there is a homotopy $H : V_0 \times I \to M$ such that $H|_{V \times \{0\}} = h_0$ and $H|_{V \times \{1\}} = h_1\varphi$ and, moreover, $H|_{V \times \{t\}} : \partial V_0 \to \partial M$ coincides with h_0. Any equivalence class of homotopy triangulation is called a *homotopy PL structure on M*. We define $\mathcal{S}_{PL}(M)$ to be the set of all homotopy PL structures on M.

If M on its own is a PL manifold, we define the *trivial element of* $\mathcal{S}_{PL}(M)$ as the equivalence class of $1_M : M \to M$.

Since every homeomorphism is a homotopy equivalence, each PL structure on M can be regarded as a homotopy triangulation. Take $M \geqslant 5$, choose a concordance class $a \in \mathcal{T}_{PL}(M)$ that is represented by a PL structure $h : V \to M$, and define $\phi(a) \in \mathcal{S}_{PL}(M)$ to be the equivalence class of the homotopy triangulation $h : V \to M$. The class $\phi(a)$ is well-defined because of Theorem 1.4.4: For $M \geqslant 5$, concordant PL structures yield equivalent homotopy triangulations. So, we obtain a function

$$\mathcal{T}_{PL}(M) \xrightarrow{\phi} \mathcal{S}_{PL}(M). \qquad (1.4.5)$$

1.4.11 Definition. Given an S^{n-1}-fibration ζ over X, a *homotopy PL prestructure on* ξ is a morphism of spherical fibrations $\varphi : \zeta \to (\gamma_{PL}^n)^\nabla$, see 1.3.12. We say that two PL prestructures $\varphi_0, \varphi_1 : \zeta \to (\gamma_{PL}^n)^\nabla$ are *equivalent* if there exists a morphism

$$\Phi : \zeta \times I \to (\gamma_{PL}^n)^\nabla$$

of spherical fibrations such that $\Phi|_{\zeta \times \{i\}} = \varphi_i, i = 0, 1$. Every such an equivalence class is called a *homotopy PL structure on* ζ.

Now, if we work for spherical fibations (not for bundles), we must use the operation † instead of $\oplus$, see 1.3.13(c). Similarly to (1.4.4), for a finite CW-space X we have a bijection

$$\lim_{n \to \infty} \{\text{homotopy PL structures on } \zeta \dagger (\theta^n)^\nabla\} \longrightarrow [X, G/PL]. \qquad (1.4.6)$$

However, here we can say more.

1.4.12 Proposition. *The sequence*

$$\{\text{homotopy PL structures on } \zeta \dagger (\theta^n)^\nabla\}_{n=1}^\infty$$

stabilizes. In particular, the map

$$\{\text{homotopy PL structures on } \zeta \dagger (\theta^n)^\nabla\} \to [G/PL]$$

is a bijection if $\dim \xi \gg \dim X$.

Proof. This follows from 1.3.19.   □

Further, for every topologically $\mathbb{R}^N$-bundle ξ that admits a PL structure we have a commutative diagram

$$
\begin{array}{ccc}
\{\text{PL structures on } \xi\} & \longrightarrow & [X, TOP/PL] \\
\downarrow & & a_* \downarrow \\
\{\text{homotopy PL structures on } \xi^\nabla\} & \longrightarrow & [X, G/PL]
\end{array}
\qquad (1.4.7)
$$

Here the right vertical map a in (1.3.7) induces the map

$$a_* : [X, TOP/PL] \to [X, G/PL].$$

This follows from Proposition 1.4.9 and its obvious analogue for G/PL. The left vertical arrow converts a morphism of $\mathbb{R}^N$-bundles into a morphism of spherical fibrations.

For a finite CW-space X, the horizontal arrows turn into bijections if we stabilize the picture, i.e., pass to the limit as in (1.4.4). Furthermore, the bottom arrow is an isomorphism if $N \gg \dim X$.

1.4.13 Remark. Actually, following the proof of the Main Theorem, one can prove that $TOP_m/PL_m = K(\mathbb{Z}/2, 3)$ for $m \geqslant 5$, see [KS2, Essay V, §5]. So, the obvious analogue of 1.3.19 holds for TOP also, and therefore the top map of the diagram (1.4.7) is a bijection for N large enough. But, of course, we are not allowed to use these a posteriori arguments here, until we accomplish the proof of the Main Theorem.

1.4.14 Remark. We can also consider *smooth* (= differentiable C^∞) structures on topological manifolds. To do this, replace the words "PL" in Definition 1.4.2 by the word "smooth". The relating set of smooth concordance classes is denoted by $\mathcal{T}_D(M)$.

The set $\mathcal{S}_D(M)$ of homotopy smooth structures is defined in the similar way: replace the words "PL" in Definition 1.4.10 by the word "smooth".

Moreover, every smooth manifold can be canonically converted into a PL manifold (S. Cairns and J. Whitehead Theorem [Cai, W], see e.g., [HM]). So, we can define the set $\mathcal{P}_D(M)$ of smooth structures on a PL manifold M. To do this, modify definition 1.4.2 as follows: M is a PL manifold with a compatible smooth boundary, V_i are smooth manifolds, h_i and H are PL isomorphisms. See [HM].

For convenience of references, we fix here the following theorem of Smale [Sma]. Actually, Smale proved it for smooth manifolds, a good proof can also be found in Milnor [Mi5]. However, the proof works for the PL case as well, cf. Stallings [Sta, 8.3, Theorem A].

1.4.15 Theorem. *Let M be a closed PL manifold that is homotopy equivalent to the sphere $S^n, n \geqslant 5$. Then M is PL homeomorphic to S^n.* □

1.4.16 Example. Now we construct an example of two smooth structures $h_i : V \to S^n, i = 1, 2$ that are not concordant. First, note that there is a bijective correspondence between $\mathcal{S}_D(S^n)$ and the Kervaire–Milnor group Θ_n of homotopy spheres, [KM]. Indeed, Θ_n consists of equivalence classes of *oriented* homotopy spheres: two oriented homotopy spheres are equivalent if they are oriented diffeomorphic (= h-cobordant). Now, given a homotopy smooth structures $h : \Sigma^n \to S^n$, we orient Σ^n so that h has degree 1. In this way we get a well-defined map $u : \mathcal{S}_D(S^n) \to \Theta_n$. Conversely, given a homotopy sphere Σ^n, consider a homotopy equivalence $h : \Sigma^n \to S^n$ of degree 1. In this way we get a well-defined map $\Theta_n \to \mathcal{S}_D(S^n)$ which is inverse to u.

Note that, because of the Smale Theorem, every smooth homotopy sphere $\Sigma^n, n \geqslant 5$, possesses a smooth function with exactly two critical points. Thus, $\mathcal{S}_D(S^n) = \mathcal{T}_D(S^n) = \mathcal{P}_D(S^n)$ for $n \geqslant 5$. Kervaire and Milnor [KM] proved that $\Theta_7 = \mathbb{Z}/28$, i.e., because of what we said above, $\mathcal{S}_D(S^7) = \mathcal{T}_D(S^7) = \mathcal{P}_D(S^7)$ consists of 28 elements.

On the other hand, there are only 15 classes of diffeomorphism of smooth manifolds which are homeomorphic (and PL homeomorphic, and homotopy equivalent) to S^7 but mutually non-diffeomorphic. Indeed, if an oriented smooth 7-dimensional manifold Σ is homeomorphic to S^7 then Σ bounds a parallelizable manifold W_Σ, [KM]. We equip W an orientation which is compatible with Σ and set

$$a(\Sigma) = \frac{\sigma(W_\Sigma)}{8} \quad \text{mod } 28$$

where $\sigma(W)$ is the signature of W. Kervaire and Milnor [KM] proved that the correspondence

$$\Theta_7 \to \mathbb{Z}/28, \quad \Sigma \mapsto a(W_\Sigma)$$

is a well-defined bijection.

However, if $a(\Sigma_1) = -a(\Sigma_2)$ then Σ_1 and Σ_2 are diffeomorphic: namely, Σ_2 is just the Σ_1 with the opposite orientation. So, there are only 15 smooth manifolds which are homotopy equivalent (and even PL homeomorphic) to S^7 but mutually non-diffeomorphic.

In terms of structures, it can be expressed as follows. Let $\rho : S^n \to S^n$ be a diffeomorphism of degree -1. Then the smooth structures $h : \Sigma^7 \to S^7$ and $\rho h : \Sigma^7 \to S^7$ are not concordant, if $a(\Sigma^7) \neq 0, 14$.

1.5   From Manifolds to Bundles

Recall that, for every topological manifold M^n, its tangent bundle τ_M and normal (with respect to an embedding $M \in \mathbb{R}^{N+n}, N \gg n$) $\mathbb{R}^N$-bundle ν_M are defined, see Milnor [Mi4]. Here τ_M is a topological $\mathbb{R}^n$-bundle, and we can regard ν_N as a topological $\mathbb{R}^N$-bundle. Here we have $\tau_M \oplus \nu_M$ is a trivial $\mathbb{R}^{N+n}$-bundle, and the total space of ν is an open subset of $\mathbb{R}^{N+n}$ (a regular neighborhood).

Furthermore, if M is a PL manifold then τ_M and ν_M turns into PL bundles in a canonical way, see [Mi3, HW]. Moreover, for N large enough, the normal bundle of a given PL embedding $M \to \mathbb{R}^{N+n}$ exists and is unique up to isomorphism. For detailed definitions and proofs, see [HW, KL, LR1].

Choose a manifold M (possibly with boundary) and a PL structure $h : V \to M$. Let $g = h^{-1} : M \to V$. Since g is a homeomorphism, it yields a topological morphism $\lambda_g : \tau_M \to \tau_V$ where τ_V, τ_M denote the tangent bundles of V, M respectively, and so we have the correcting topological morphism $c(\lambda_g) : \tau_M \to \lambda^* \tau_V$. Let $\nu = \nu_M^N$ be a normal bundle of M in $\mathbb{R}^{N+n}$ with N large enough.

Now, assume that M on its own is (a priori) a PL manifold. Then $\nu = \nu_M^N$ turns out to be a PL bundle. Consider the topological morphism

$$\mu = \mu(h) : \theta_M^{N+n} = \tau_M \oplus \nu_M^N \longrightarrow g^* \tau_V \oplus \nu_M^N \xrightarrow{\text{classif}} \gamma_{PL}^{N+n}$$

and regard it as a PL structure on θ^{N+n}.

1.5.1 Definition. Define a map

$$j_{TOP} : \mathcal{T}_{PL}(M) \to [M, TOP/PL]$$

as the composition

$$\mathcal{T}_{PL}(M) \longrightarrow \lim_{N \to \infty} \{\text{PL structures on } \theta_M^N\} \longrightarrow [M, TOP/PL]$$

where the first map assigns the morphism μ to the structure h, and the last map comes from (1.4.4). The map j_{TOP} is well-defined because concordant PL structures on M yield the same PL structure on its tangent bundle.

Now we construct a map $j_G : \mathcal{S}_{PL}(M) \to [M, G/PL]$, a "homotopy analogue" of j_{TOP}. This construction is more delicate, and we treat here the case of closed manifolds only. So, let M be a connected closed PL manifold.

Roughly (informally) speaking, we do the following. Let $h : V \to M$ represent a class $x \in \mathcal{S}_{PL}$. Let $g : M \to V$ be a homotopy inverse to h. Consider an embedding $M \to V \times \mathbb{R}^N$ such that the map

$$M \longrightarrow V \times \mathbb{R}^N \xrightarrow{\text{projection}} V$$

is homotopic to g. Let ν be a normal bundle of M in $V \times \mathbb{R}^N$, and let $f : M \to BPL$ classify ν. Then the map

$$M \xrightarrow{f} BPL \xrightarrow{\alpha} BG$$

is inessential because g is a homotopy equivalence. So, f lifts to G/PL, i.e., we have a lifting $\tilde{f} : M \to G/PL$. Now we define $j_G(x)$ as the homotopy class of $\tilde{f}$.

Now we pass to more accurate exposition.

First, note that the TOP_n-action on $\mathbb{R}^n$ extends uniquely to a TOP_n-action on the one-point compactification S^n of $\mathbb{R}^n$. Now, given a topological (or PL) $\mathbb{R}^n$ bundle, we regard ξ as a $(TOP_n, \mathbb{R}^n)$-bundle and put $\xi^\bullet = \{p^\bullet : E^\bullet \to B\}$ to be the (TOP_n, S^n)-bundle with respect to the above TOP_n-action on S^n. Note that the fixed point "infinity" of the TOP_n-action on S^n yeilds a section $s^\bullet : B \to E^\bullet \setminus E$.

Moris Hirsch [H, Theorem B] proved the following uniqueness claim: If you have a spherical sectioned bundle ξ' over B such that ξ is the interior of ξ' then ξ' and $\xi^\bullet$ are isomorphic.

1.5.2 Definition. (a) Given a topological $\mathbb{R}^N$-bundle $\xi = \{p : E \to B\}$, we define the Thom space $T\xi$ as $T\xi = E^\bullet/s^\bullet(B)$.

(b) Given a spherical fibration $\eta = \{q : Y \to B\}$, we define the Thom space $T\eta$ as the cone $C(q)$ of the map $g : Y \to B$.

(c) To connect (a) and (b), note that $T(\xi^\triangledown)$ and $T\xi$ are homotopy equivalent (see 1.3.12 for notation $\xi^\triangledown$).

1.5.3 Construction–Definition. We embed M^n in $\mathbb{R}^{N+n}$ where $N \gg n$ and let $\nu = \nu_M, \dim \nu = N$ be a normal bundle of the embedding. Recall that ν is a PL bundle $E \to M$ whose total space E is PL homeomorphic to a (regular) neighborhood U of M in S^{N+n}. We choose such isomorphism and denote it by $\varphi : U \to E$.

Let $T\nu$ be the Thom space of ν. Then there is a unique map

$$\psi : \mathbb{R}^{N+n}/(\mathbb{R}^{N+n} \setminus U) \to T\nu$$

such that $\psi|_U = \varphi$. We regard S^{N+n} as the one-point compactification of $\mathbb{R}^{N+n}$ and define the *collapsing map* $\iota : S^{N+n} \to T\nu_M$ (the *Browder–Novikov map*, cf. [Br1,N1]) to be the composition

$$\iota : S^{N+n} \xrightarrow{\text{quotient}} S^{N+n}/(S^{N+n} \setminus U) = \mathbb{R}^{N+n}/(\mathbb{R}^{N+n} \setminus U) \xrightarrow{\psi} T\nu.$$

1.5.4 Remark. The uniqueness of the normal bundle $\nu^N, N \gg n$ yields the following important fact. Let $\nu' = \{E' \to M\}$ be another normal bundle and $\varphi' : U' \to E'$ be another PL homeomorphism. Let $\iota : S^{N+n} \to T\nu$ and $\iota' : S^{N+n} \to T\nu'$ be the corresponding Browder–Novikov collapsing maps. Then there is a morphism $\nu \to \nu'$ of PL bundles which carries ι to a map homotopic to ι'.

1.5.5 Definition. A pointed space X is called *reducible* if there is a pointed map $f : S^m \to X$ such that $f_* : \tilde{H}_i(S^m) \to \tilde{H}_i(X)$ is an isomorphism for $i \geqslant m$. Every such map f (as well as its homotopy class or its stable homotopy class) is called a *reducibility* for X.

Geometric intuition hints that collapsing map ι in 1.5.3 is a reducibility for $T\nu$, and this is really true, see Corollary 2.2.3 below.

Now, consider data as in 1.5.3. Recall that ν^∇ denotes the bundle ν with deleted zero section, see 1.3.12. Note also that we can regard an element $\iota \in \pi_{N+n}(T\nu)$ as $\iota \in \pi_{N+n}(T\nu^\nabla)$, cf. 1.5.2.

1.5.6 Theorem. *Let η be an S^{N-1}-fibration over M such that $T\eta$ is reducible. Let $\alpha \in \pi_{N+n}(T\eta)$ be an arbitrary reducibility for $T\eta$. Then there exists a fiberwise S^{N-1}-equivalence $\mu : \nu^\nabla \to \eta$ such that $(T\mu)_*(\iota) = \alpha$, and such μ is unique up to fiberwise homotopy over M.*

This is a version of a theorem of Wall [W2, Theorem 3.5], cf. also [Spi, Br1, N1]. We postpone the proof to the next Chapter, see Section 2.2.

Given a closed connected PL manifold M, consider a homotopy triangulation $h : V \to M$ and let ν_V be a normal bundle of a certain embedding $V \subset \mathbb{R}^{N+n}$, and let $u \in \pi_{N+n}(T\nu_V)$ be the homotopy class of the collapsing map $S^{N+n} \to T\nu_V$. Let $g : M \to V$ be a homotopy inverse to h and set $\eta = g^* \nu_V^\nabla$. The g-adjoint morphism

$$\mathfrak{I} = \mathfrak{I}_g : \eta = g^* \nu_V^\nabla \to \nu_V^\nabla$$

yields the map $T\mathfrak{I} : T\eta \to T\nu_V^\nabla$. It is easy to see that $T\mathfrak{I}$ is a homotopy equivalence, and so there exists a unique $\alpha \in \pi_{N+n}(T\eta)$ with $(T\mathfrak{I})_*(\alpha) = u$. Since u is a reducibility for $T\nu_V$, we conclude that α is a reducibility for $T\eta$. By Theorem 1.5.6, we get an S^{N-1}-equivalence $\mu : \nu^\nabla \to \eta$ with $(T\mu)_*(\iota) = \alpha$ and $\nu = \nu_M$. Now, the morphism

$$\nu^\nabla \xrightarrow{\ \mu\ } \eta \xrightarrow{\ \text{classif}\ } \gamma_G^N \tag{1.5.1}$$

is a homotopy PL prestructure on ν^∇. Because of the uniqueness of the normal bundle, the homotopy class of this prestructure is well-defined. So, we assigned a homotopy PL structure on ν^∇ to a homotopy triangulation. The homotopic homotopy triangulations yield the same structure on ν^∇, and we have the well-defined function

$$\mathcal{S}_{PL}(M) \longrightarrow \{\text{homotopy PL structures on } \nu_M^\nabla\}.$$

1.5.7 Definition. The *normal invariant* is the function
$$j_G : \mathcal{S}_{PL}(M) \to [M, G/PL]$$
that is defined as the composition
$$\mathcal{S}_{PL}(M) \longrightarrow \{\text{homotopy PL structures on } \nu_M^\triangledown\} \cong [M, G/PL]$$
where the bijection at the right comes from 1.4.12. The value of j_G on a homotopy PL structure (as well as on its homotopy triangulation) is called the normal invariant of the structure.

1.6 Homotopy PL Structures on $T^n \times D^k$

Below T^n denotes the n-dimensional torus.

1.6.1 Theorem. *Assume that $k + n \geqslant 5$. If $x \in \mathcal{S}_{PL}(T^n \times S^k)$ can be represented by a homeomorphism $M \to T^n \times S^k$ then $j_G(x) = 0$.* $\square$

We prove 1.6.1 (in fact, a little bit more general result) in the next chapter.

We also prove more general fact, the Sullivan Normal Invariant Homeomorphism Theorem 3.3.2. It claims that $j_F(h) = 0$ provided that $h : V \to M$ is a homeomorphism, $\dim M \geqslant 5$, and $H_3(M)$ is 2-torsion free. We need to repeat this (make a loop, first prove the special case (Theorem 1.6.1 of 3.3.2) separately, then prove 3.3.2) in order to avoid the logic circle, since the full proof of Theorem 3.3.2 uses Main Theorem and hence Theorem 1.6.1.

1.6.2 Construction–Definition. Let $x \in \mathcal{S}_{PL}(M)$ be represented by a homotopy triangulation $h : V \to M$, and let $p : \widetilde{M} \to M$ be a covering. Then we have a commutative diagram

$$
\begin{array}{ccc}
\widetilde{V} & \xrightarrow{\ \widetilde{h}\ } & \widetilde{M} \\
{\scriptstyle q}\downarrow & & \downarrow{\scriptstyle p} \\
V & \xrightarrow{\ h\ } & M
\end{array}
$$

where q is the induced covering. Since $\widetilde{h}$ is defined by h uniquely up to deck transformations, the equivalence class of $\widetilde{h}$ is well defined. Moreover, if we vary h in its equivalence class x then $\widetilde{h}$ varies in its equivalence class in $\mathcal{S}_{PL}(\widetilde{M})$. So, we have a well-defined map
$$p^* : \mathcal{S}_{PL}(M) \to \mathcal{S}_{PL}(\widetilde{M})$$
where $p^*(x)$ is the equivalence class of $\widetilde{h}$.

If p is a finite covering, we say that a class $p^*(x) \in \mathcal{S}_{PL}(\widetilde{M})$ *finitely covers* the class x.

1.6.3 Theorem (Hsiang–Shaneson [HS], Wall [W4]). *Let $k + n \geqslant 5$. Then the following holds:*

(i) *if $k > 3$ then the set $\mathcal{S}_{PL}(T^n \times D^k)$ consists of precisely one (trivial) element;*

(ii) *if $k < 3$ then every element of $\mathcal{S}_{PL}(T^n \times D^k)$ can be finitely covered by the trivial element;*

(iii) *the set $\mathcal{S}_{PL}(T^n \times D^3)$ contains at most one element which cannot be finitely covered by the trivial element.*

Some words about the proof. First, we mention the proof given by Wall, [W3] and [W4, Section 15 A]. Wall proved the bijection

$$w : \mathcal{S}_{PL}(T^n \times D^k) \to H^{3-k}(T^n).$$

Moreover, he also proved that finite coverings respect the bijection, i.e., if $p : T^n \times D^k \to T^n \times D^k$ is a finite covering then the commutative diagram

$$
\begin{array}{ccc}
\mathcal{S}_{PL}(T^n \times D^k) & \xrightarrow{\ w\ } & H^{3-k}(T^n; \mathbb{Z}/2) \\
{\scriptstyle p^*}\big\uparrow & & \big\uparrow{\scriptstyle p^*} \\
\mathcal{S}_{PL}(T^n \times D^k) & \xrightarrow{\ w\ } & H^{3-k}(T^n; \mathbb{Z}/2)
\end{array}
$$

commutes. Certainly, this result implies the assertions (i)–(iii). Wall's proof uses difficult algebraic calculations.

Another and more geometric proof of the theorem can be found in the paper of Hsiang and Shaneson [HS, Theorem C]. We do not want to copy the proof, but we emphasize that the inequality

$$\sigma_{PL}^1 = 16 \neq 8 = \sigma_{TOP}^1$$

(see p. xiv) gives rise to the singular role of the number $3 = \dim D^3$ in 1.6.3(iii).

This is worthy to say that the proof in [HS] uses Theorem 1.6.1. Indeed, Hsiang and Shaneson consider the so-called surgery exact sequence

$$\xrightarrow{\ \partial\ } \mathcal{S}_{PL}(S^k \times T^n) \xrightarrow{\ j_G\ } [S^k \times T^n, G/PL] \to \qquad (1.6.1)$$

and write (p. 42, Section 10):

By [44], every homomorphism

$$h : M \to S^k \times T^n, k + n \geq 5,$$

represents an element in the image of ∂.

(Here bibliographic item [44] from the citation is our bibliographical item [Sul1].) In other words, because of the exactness of sequence (1.6.1), we have $j_G(h) = 0$ whenever $h : M \to S^k \times T^n$ is a homeomorphism. This is exactly Theorem 1.6.1. As I already wrote, you will see the proof in Chapter 2.   □

1.7 The Product Structure Theorem, or from Bundles to Manifolds

Let M be an n-dimensional ∂_{PL}-manifold. Then every PL structure $h : V \to M$ yields a PL structure

$$h \times 1 : V \times \mathbb{R}^k \to M \times \mathbb{R}^k.$$

Thus, we have a well-defined map

$$e : \mathcal{T}_{PL}(M) \to \mathcal{T}_{PL}(M \times \mathbb{R}^k), \quad h \mapsto h \times 1.$$

1.7.1 Theorem (The Product Structure Theorem). *For every $n \geqslant 5$ and every $k \geqslant 0$, the map $e : \mathcal{T}_{PL}(M) \to \mathcal{T}_{PL}(M \times \mathbb{R}^k)$ is a bijection. In particular, if $\mathcal{T}_{PL}(M \times \mathbb{R}^k) \neq \emptyset$ then $\mathcal{T}_{PL}(M) \neq \emptyset$.*

Kirby and Siebenmann made the breakthrough for Theorem 1.7.1, see [KS1, KS2, K1]. Quinn [Q4] gave a relative short proof of 1.7.1 by developing his deep and difficult theory of ends of maps, [Q1, Q2]. We outline these ideas in Appendix.

1.7.2 Corollary (The Classification Theorem). *If $\dim M \geqslant 5$ and M admits a PL structure, then the map*

$$j_{TOP} : \mathcal{T}_{PL}(M) \to [M, TOP/PL]$$

is a bijection.

Proof. We construct a map

$$\sigma : [M, TOP/PL] \to \mathcal{T}_{PL}(M) \tag{1.7.1}$$

which is inverse to j_{TOP}. Take an element

$$a \in [M, TOP/PL]$$

and, using (1.4.4), interpret it as a homotopy class of a topological $\mathbb{R}^N$-morphism $\varphi : \theta_M^N \to \gamma_{PL}^N$ such that $\varphi|_{\partial M}$ is a PL $\mathbb{R}^N$-morphism. The morphism φ yields a correcting isomorphism $\theta_M^N \to b^*\gamma_{PL}^N$ of topological

$\mathbb{R}^N$-bundles over M, where $b : M \to BPL$ is the base of the morphism φ. So, we have the commutative diagram

where h is a fiberwise homeomorphism and $W \to M$ is a PL $\mathbb{R}^N$-bundle $b^*\gamma_{PL}^N$. In particular, W is a PL manifold. Regarding $h^{-1} : W \to M \times \mathbb{R}^N$ as a PL structure on $M \times \mathbb{R}^N$, we conclude that, by the Product Structure Theorem 1.7.1, h^{-1} is concordant to a map $g \times 1$ for some PL structure $g : V \to M$. We define $\sigma(a) \in \mathcal{T}_{PL}(M)$ to be the concordance class of g. One can check that σ is a well-defined map which is inverse to j_{TOP}. Cf. [KS2, Essay IV, Theorem 4.1]. $\qquad\square$

1.7.3 Remark. Concerning dimension 4. (Note that there is no difference between PL and smooth case, see 1.7.8.) The concordance classes of PL structures on a connected non-compact 4-manifold M correspond bijectively with the elements of $H^3(M; \mathbb{Z}/2)$, [FQ, §8]. However, unlike case $\dim M \geqslant 5$, we can't say that the *Hauptvermutung* holds for M whenever $H^3(M; \mathbb{Z}/2) = 0$. Indeed, there are infinite family PL structures on $\mathbb{R}^4$ that are mutually non-PL-homeomorphic.

1.7.4 Corollary (The Existence Theorem). *A topological manifold M with $\dim M \geqslant 5$ admits a PL structure if and only if the tangent bundle of M admits a PL structure.*

Proof. We already noticed (in Section 1.5) that any PL manifold admits a tangent PL bundle. This proves "only if" parts. To prove "if" part, let $\tau = \{\pi : D \to M\}$ be the tangent bundle of M, and let $\nu = \{r : E \to M\}$ be a stable normal bundle of M, $\dim \nu = N$. Then E is homeomorphic to an open subset of $\mathbb{R}^{N+n}$, and therefore we can (and shall) regard E as a PL manifold. Since τ is a PL bundle, we conclude that $r^*\tau$ is a PL bundle over E. In particular, the total space $M \times \mathbb{R}^{N+n}$ of $r^*\tau$ turns out to be a PL manifold, cf. [Mi4]. Now, because of the Product Structure Theorem 1.7.1, M admits a PL structure. Cf. [KS2, Essay IV, Theorem 3.1]. $\qquad\square$

Let $f : M \to BTOP$ classify the stable tangent bundle of a closed topological manifold M, $\dim M \geqslant 5$.

1.7.5 Corollary. *The following conditions are equivalent:*
  (i) *M admits a PL structure;*
  (ii) *τ admits a PL structure;*
  (iii) *there exists k such that $\tau \oplus \theta^k$ admits a PL structure;*
  (iv) *the map f admits an α_{TOP}^{PL}-lifting to BPL.*

Proof. The implications (i) $\Longrightarrow$ (iv) and (ii) $\Longrightarrow$ (iii) are obvious . So, it
suffices to prove that (iv) $\Longrightarrow$ (iii) $\Longrightarrow$ (i). The implication (iii) $\Longrightarrow$ (i) can
be proved similarly to 1.7.4. Furthermore, since M is compact, we conclude
that $f(M) \subset BTOP_m$ for some m. So, if (iv) holds then f lifts to BPL_m,
i.e., $\tau \oplus \theta^{m-k}$ admits a PL structure.   □

1.7.6 Remark. It follows from 1.2.4, 1.7.4, and 1.7.2 that the set $\mathcal{T}_{PL}(M)$
of concordance classes of PL structures on M correspond bijectively with
the set of fiber homotopy classes of α_{TOP}^{PL}-liftings of f.

1.7.7 Remark. It is well known that j_G is not a bijection in general. The
"kernel"' and "cokernel" of j_G can be described in terms of so-called Wall
groups, [W4]. (For M simply-connected, see also Theorem 2.4.2.) On the
other hand, the bijectivity of j_{TOP} (the Classification Theorem) follows
from the Product Structure Theorem. So, informally speaking, kernel and
cokernel of j_G play the role of obstructions to splitting of structures. It
seems interesting to develop and clarify these naive arguments.

1.7.8 Remark. Since tangent and normal bundles of smooth manifolds turn out to be
vector bundles, one can construct a map
$$k : \mathcal{P}_D(M) \to [M, PL/O]$$
which is an obvious analogue of j_{TOP}. Moreover, the obvious analogue of the Product Structure Theorem (as well as of the Classification and Existence Theorems) holds
without any dimensional restriction. In particular, k is a bijection for all smooth manifolds, [HM].
    It is well known (although difficult to prove) that $\pi_i(PL/O) = 0$ for $i \leqslant 6$. (See [Rud],
IV.4.27(iv)] for the references.) Thus, every PL manifold M of dimension $\leqslant 7$ admits a
smooth structure, and this structure is unique (up to diffeomorphism) if $\dim M \leqslant 6$.

1.8   Non-contractibility of TOP/PL

1.8.1 Theorem (Rokhlin Signature Theorem). *Let M be a closed 4-
dimensional PL manifold with $w_1(M) = 0 = w_2(M)$. Then the signature
of M is divisible by 16.*

Proof. See [MK], [K2, Chapter XI], or the original work [Ro]. In fact,
Rokhlin proved the result for smooth manifolds, but the proof works for

PL manifolds as well. On the other hand, in view of 1.7.8, there is no difference between smooth and PL manifolds in dimension 4. □

1.8.2 Theorem (Freedman's Example). *There exists a closed simply-connected topological 4-dimensional manifold V with $w_2(V) = 0$ and such that E_8 is the matrix of the intersection form $H_2(V) \otimes H_2(V) \to \mathbb{Z}$. In particular, the signature of V is equal to 8. Furthermore, such a manifold V is unique up to homeomorphism.*

Proof. See [FU], [FQ], or the original work [F]. □

1.8.3 Comment. Some words on constructing of V. Take the manifold W (Milnor's plumbing) described in [Br2, Complement V.2.6]. This is a smooth 4-dimensional simply-connected parallelizable manifold whose boundary ∂W is a homology sphere. Furthermore, E_8 is the matrix of the intersection form $H_2(W) \otimes H_2(W) \to \mathbb{Z}$. A key (and very difficult) result of Freedman [F] states that ∂W bounds a compact contractible topological manifold P. Now, put $V = W \cup_{\partial W} P$.

1.8.4 Corollary. *The topological manifolds V and $V \times T^k, k \geqslant 1$ do not admit any PL structure.*

Proof. The statement about V follows from 1.8.1. (Note that $w_1(V) = 0$ because V is simply-connected.) Suppose that $V \times T^k$ admits a PL structure. Then $V \times \mathbb{R}^k$ admits a PL structure. So, because of the Product Structure Theorem 1.7.1, $V \times \mathbb{R}$ admits a PL structure. Hence, because of 1.7.8, we can assume that $V \times \mathbb{R}$ possesses a smooth structure. Then the projection $p_2 : V \times \mathbb{R} \to \mathbb{R}$ can be C^0-approximated by a map $f : V \times \mathbb{R} \to \mathbb{R}$ which coincides with p_2 on $V \times (-\infty, 0)$ and is smooth on $V \times (1, \infty)$. Take a regular value $a \in (1, \infty)$ of f (which exists because of the Sard Theorem) and set $U = f^{-1}(a)$. Then U is a smooth manifold (by the Implicit Function Theorem), and it is easy to see that $w_1(U) = 0 = w_2(U)$ (because it holds for both manifolds $\mathbb{R}$ and $V \times \mathbb{R}$). On the other hand, both manifolds V and U cut the "tube" $V \times \mathbb{R}$. So, they are (topologically) bordant, and therefore U has signature 8. But this contradicts the Rokhlin Theorem 1.8.1. □

1.8.5 Corollary. *The space TOP/PL is not contractible.*

Proof. Indeed, suppose that TOP/PL is contractible. Then every map $X \to BTOP$ lifts to BPL, and so, by 1.7.4, every closed topological manifold of dimension greater than 4 admits a PL structure. But this contradicts 1.8.4. $\qquad\Box$

1.8.6 Remark. Siebenmann [Sieb4] constructed the original example of a topological manifold (in dimensional 5) which does not admit a PL structure. Namely, consider the space

$$X^4 = T^4 \# (W \cup \text{ cone over } \partial W)$$

with W as in 1.8.3. We can't claim that X is a topological manifold, but it can be proved that $X \times S^1$ is. If we assume that $X \times S^1$ admits a PL structure, and argue as in the final part of the proof of 1.8.4, we construct a manifold Y (an analog of U in 1.8.4) with $w_1(Y) = 0 = w_2(Y)$ and $\sigma(Y) = 8$. This contradicts the Rokhlin Theorem. Thus, the 5-manifold $X \times S^1$ does not admit a PL structure.

So, you see that this argument is parallel to that of Corollary 1.8.4. From this point of view, Freedman replaced the cone over ∂W by a contractible manifold P in 1.8.3. This allowed Freedman to construct a a 4-dimensional manifold (namely, V) that is not PL.

1.9   Homotopy Groups of TOP/PL

Let M be a compact topological manifold equipped with a metric ρ. Then the space $\mathscr{H}$ of self-homeomorphisms $M \to M$ gets a metric d with $d(f,g) = \sup\{x \in M \mid \rho(f(x), g(x))\}$.

1.9.1 Theorem. *The space $\mathscr{H}$ is locally contractible.*

Proof. See [Ch, EK]. $\qquad\Box$

1.9.2 Corollary. *There exists $\varepsilon > 0$ such that every homeomorphism $h \in \mathscr{H}$ with $d(H, 1_M) < \varepsilon$ is isotopic to 1_M.* $\qquad\Box$

1.9.3 Construction. We regard the torus T^n as a commutative Lie group (multiplicative) and equip it with the invariant metric ρ. Consider the map $p_\lambda : T^n \to T^n, p_\lambda(a) = a^\lambda, \lambda \in \mathbb{N}$. Then p_λ is a λ^n-sheeted covering. It is also clear that all the deck transformations of the covering torus are isometries. Hence the diameter of each of (isometric) fundamental domain for p_λ tends to zero as $\lambda \to \infty$.

1.9.4 Lemma. *Let $h : T^n \times D^k \to T^n \times D^k$ be a self-homeomorphism which is homotopic rel $\partial(T^n \times D^k)$ to the identity. Then there exist $\lambda \in \mathbb{N}$*

and a commutative diagram

$$
\begin{array}{ccc}
T^n \times D^k & \xrightarrow{\ \widetilde{h}\ } & T^n \times D^k \\
{\scriptstyle p_\lambda}\Big\downarrow & & \Big\downarrow{\scriptstyle p_\lambda} \\
T^n \times D^k & \xrightarrow{\ h\ } & T^n \times D^k
\end{array}
$$

where the lifting $\widetilde{h}$ of h is isotopic rel $\partial(T^n \times D^k)$ to the identity.

Proof. (Cf. [KS2, Essay V].) First, consider the case $k = 0$. Without loss of generality we can assume that $h(e) = e$ where e is the neutral element of T^n. Consider a covering $p_\lambda : T^n \to T^n$ as in 1.9.3 and take a covering $\widetilde{h} : T^n \to T^n, p_\lambda \widetilde{h} = h p_\lambda$ of h such that $\widetilde{h}(e) = e$. In order to distinguish the domain and the range of p_λ, we denote the domain of p_λ by $\widetilde{T}^n$ and the range of p_λ by T^n. Since h is homotopic to 1_T, we conclude that every point of the lattice $L := p_\lambda^{-1}(e)$ is fixed under $\widetilde{h}$.

Given $\varepsilon > 0$, choose δ such that $\rho(\widetilde{h}(x), \widetilde{h}(y)) < \varepsilon/2$ whenever $\rho(x,y) < \delta$. Furthermore, choose λ so large that the diameter of any closed fundamental domain is less than $\min\{\varepsilon/2, \delta\}$. Now, given $x \in \widetilde{T}$, choose $a \in L$ such that a and x belong to the same closed fundamental domain. Now,

$$
\rho(x, \widetilde{h}(x)) \leqslant \rho(x,a) + \rho(a, \widetilde{h}(x)) = \rho(x,a) + \rho(\widetilde{h}(a), \widetilde{h}(x))
$$
$$
< \frac{\varepsilon}{2} + \frac{\varepsilon}{2} = \varepsilon.
$$

So, for every $\varepsilon > 0$ there exists λ such that $d(\widetilde{h}, 1_{\widetilde{T}}) < \varepsilon$. Thus, by 1.9.2, $\widetilde{h}$ is isotopic to $1_{\widetilde{T}}$ for λ large enough.

The proof for $k > 0$ is similar but a bit more technical. Let $D_\eta \subset D^k$ be the disk centered at 0 and having the radius η. We can always assume that h coincides with identity outside of $T^n \times D_\eta$. Now, asserting as for $k = 0$, take a covering p_λ as above and choose λ and η so small that the diameter of every fundamental domain in $\widetilde{T}^n \times D_\eta$ is small enough. Then

$$
\widetilde{h} : \widetilde{T} \times D_\eta \to \widetilde{T} \times D_\eta
$$

is isotopic to the identity and coincides with identity outside $\widetilde{T} \times D_\eta$. This isotopy is not an isotopy rel $\widetilde{T} \times \partial D_\eta$. Nevertheless, we can easily extend it to the whole $\widetilde{T} \times D^k$ so that this extended isotopy is an isotopy rel $\partial(\widetilde{T} \times D^k)$.

If you want an explicit formula, do the following. Given $a = (b,c) \in \widetilde{T} \times D_\eta$, set $|a| = |c|$. Consider an isotopy

$$
\varphi : \widetilde{T} \times D_\eta \times I \to \widetilde{T} \times D_\eta \times I, \quad \varphi(a,0) = a, \ \varphi(a,1) = \widetilde{h}(a), \quad a \in \widetilde{T} \times D_\eta.
$$

Define $\overline{\varphi} : \widetilde{T} \times D_\eta \times I \to \widetilde{T} \times D_\eta \times I$ by setting

$$\overline{\varphi}(a,t) = \begin{cases} \varphi(a,t) & \text{if } |a| \leqslant \eta, \\ \varphi(a, \frac{|a|-1}{\eta-1}t) & \text{if } |a| \geqslant \eta. \end{cases}$$

Then $\overline{\varphi}$ is the desired isotopy rel $\partial(\widetilde{T} \times D^k)$. $\square$

1.9.5 Corollary. *Let* $\phi : \mathcal{T}_{PL}(T^n \times D^k) \to \mathcal{S}_{PL}(T^n \times D^k)$ *be the forgetful map as in* (1.4.5). *If* $\phi(x) = \phi(y)$ *then there exists a finite covering* $p : T^n \times D^k \to T^n \times D^k$ *such that* $p^*(x) = p^*(y)$. $\square$

Let $p_2 : T^n \times D^k \to D^k$ be the projection. Consider the map

$$\psi : \pi_k(TOP/PL) = [(D^k, \partial D^k), (TOP/PL, *)] \xrightarrow{p_2^*}$$
$$[(T^n \times D^k, \partial(T^n \times D^k)), (TOP/PL, *)] \xrightarrow{\sigma} \mathcal{T}_{PL}(T^n \times D^k)$$

where σ is the map from (1.7.1) (the inverse map to j_{TOP}).

1.9.6 Lemma. *The map* ψ *is injective. Moreover, if, for some finite covering*

$$p : T^n \times D^k \to T^n \times D^k$$

we have $p^*\psi(x) = p^*\psi(y)$ *then* $x = y$. *In particular, if* $p^*\psi(x)$ *is the trivial element of* $\mathcal{T}_{PL}(T^n \times D^k)$ *then* $x = 0$.

Proof. The injectivity of ψ follows from the injectivity of p_2^* and σ. Furthermore, for every finite covering $p : T^n \times D^k \to T^n \times D^k$ we have the commutative diagram

$$
\begin{array}{ccc}
\pi_k(TOP/PL) & \xrightarrow{\psi} & \mathcal{T}_{PL}(T^n \times D^k) \\
\Big\| & & \Big\uparrow{\scriptstyle p^*} \\
\pi_k(TOP/PL) & \xrightarrow{\psi} & \mathcal{T}_{PL}(T^n \times D^k)
\end{array}
$$

Therefore $x = y$ whenever $p^*\psi(x) = p^*\psi(y)$. Finally, if $p^*\psi(x)$ is trivial element then $p^*\psi(x) = p^*\psi(0)$, and thus $x = 0$. $\square$

Consider the map

$$\varphi : \pi_k(TOP/PL) \xrightarrow{\psi} \mathcal{T}_{PL}((T^n \times D^k) \xrightarrow{\phi} \mathcal{S}_{PL}((T^n \times D^k)$$

where ϕ is the forgetful map described in (1.4.5).

1.9.7 Theorem (The Reduction Theorem, cf. [K1]). *The map φ is injective. Moreover, if, for some finite covering*

$$p : T^n \times D^k \to T^n \times D^k$$

we have $p^*\varphi(x) = p^*\varphi(y)$ then $x = y$. In particular, if $p^*\varphi(x)$ is the trivial element of $\mathcal{T}_{PL}(T^n \times D^k)$ then $x = 0$.

We call it the Reduction Theorem because it *reduces* the evaluation of the group $\pi_k(TOP/PL)$ to the evaluation of the sets $\mathcal{S}_{PL}(T^n \times D^k)$.

Proof. If $\varphi(x) = \varphi(y)$ then $\phi(\psi(x)) = \phi(\psi(y))$. Hence, by Corollary 1.9.5, there exists a finite covering $\pi : T^n \times D^k \to T^n \times D^k$ such that $\pi^*\psi(x) = \pi^*\psi(y)$. So, by Lemma 1.9.6, $x = y$, i.e., φ is injective.

Now, suppose that $p^*\varphi(x) = p^*\varphi(y)$ for some finite covering

$$p : T^n \times D^k \to T^n \times D^k.$$

Then $\phi(p^*\psi(x)) = \phi(p^*\psi(y))$. Now, by Corollary 1.9.5, there exists a finite covering

$$q : T^n \times D^k \to T^n \times D^k$$

such that $q^*p^*\psi(x) = q^*p^*\psi(y)$, i.e., $(pq)^*\psi(x) = (pq)^*\psi(y)$. Thus, $x = y$ by Lemma 1.9.6. □

1.9.8 Corollary. *We have $\pi_i(TOP/PL) = 0$ for $i \neq 3$. Furthermore, $\pi_3(TOP/PL)$ is a subgroup of the group $\mathbb{Z}/2$. So, $TOP/PL = K(\pi, 3)$ where $\pi = \mathbb{Z}/2$ or $\pi = 0$.*

Proof. The equality $\pi_i(TOP/PL) = 0$ for $i \neq 3$ follows from Theorem 1.6.3(i,ii) and Theorem 1.9.7. Furthermore, again because of 1.6.3 and 1.9.7, we conclude that $\pi_3(TOP/PL)$ has at most two elements. In other words, $TOP/PL = K(\pi, 3)$ where $\pi = \mathbb{Z}/2$ or $\pi = 0$. □

Now, the Main Theorem.

1.9.9 Theorem. *We have $\pi_i(TOP/PL) = 0$ for $i \neq 3$. Furthermore, $\pi_3(TOP/PL) = \mathbb{Z}/2$. Thus, there is a homotopy equivalence $TOP/PL \simeq K(\mathbb{Z}/2, 3)$.*

Proof. We see from Corollary 1.9.8 that $TOP/PL = K(\pi, 3)$ where $\pi = \mathbb{Z}/2$ or $\pi = 0$. Furthermore, the space TOP/PL is not contractible by Corollary 1.8.5. Thus, $TOP/PL \simeq K(\mathbb{Z}/2, 3)$. □

Chapter 2

Normal Invariant

The purpose of this chapter is to prove Theorem 1.6.1. The proof uses the Sullivan's result on the homotopy type of G/PL, [Sul1, Sul2]. Note that Madsen and Milgram [MM] gave a detailed proof of this Sullivan's result.

2.1 Preliminaries on Stable Duality

We need certain information concerning stable duality [Hus, Spa1, Sw]. Given a pointed map $f : X \to Y$, let $Sf : SX \to SY$ denote the reduced suspension over f. So, we have a well-defined map

$$S : [X,Y]^\bullet \to [SX, SY]^\bullet, \quad [f] \to [Sf].$$

2.1.1 Theorem. *Suppose that $\pi_i(Y) = 0$ for $i < n$ and that X is a CW-space with $\dim X < 2n - 1$. Then the map $S : [X,Y]^\bullet \to [SX, SY]^\bullet$ is a bijection.*

Proof. This is the famous Freudenthal Suspension Theorem, see e.g., [FFG, Hat, Spa2, Sw, Wh2] □

2.1.2 Notation. Given two pointed spaces X, Y, we define $\{X, Y\}$ to be the direct limit of the sequence

$$[X,Y]^\bullet \xrightarrow{S} [SX, SY]^\bullet \xrightarrow{S} \cdots \longrightarrow [S^n X, S^n Y]^\bullet \xrightarrow{S} \cdots .$$

In particular, we have the obvious maps

$$\Sigma : (Y, *)^{(X,*)} \longrightarrow [X,Y]^\bullet \longrightarrow \{X, Y\}.$$

2.1.3 Definition. Given a pointed map $f : X \to Y$, the element $\Sigma(f) \in \{X, Y\}$ is called the *stable homotopy class of f*. The standard notation for this one is $\{f\}$, but, as usual, in several cases we use the same notation f for f, $[f]$, and $\{f\}$.

It is well known that, for $n \geqslant 2$, the set $[S^n X, S^n Y]^\bullet$ has a natural structure on the Abelian group, and the corresponding maps S are homomorphisms, [Sw, Wh2]. So, $\{X, Y\}$ turns out to be a group. Furthermore, by Theorem 2.1.1, if X is a finite CW-space then the map

$$[S^N X, S^N Y]^\bullet \to \{S^N X, S^N Y\}$$

is a bijection for N large enough.

2.1.4 Proposition-Definition. (a) A map $f : S^q \to A \wedge A^\perp$ is called *a q-duality* if, for every space E, the maps

$$u_E : \{A, E\} \to \{S^q, E \wedge A^\perp\}, \quad u_E(\varphi) = (\varphi \wedge 1_{A^\perp})u$$

and

$$u^E : \{A^\perp, E\} \to \{S^q, A \wedge E\}, \quad u^E(\varphi) = (1_A \wedge \varphi)u$$

are isomorphisms.

Frequently people call the map f *stable q-duality*, but we will not discuss the unstable duality and therefore omit the adjective "stable".

(b) We say that a space $A^\perp$ is q-dual to A if there is a q-duality $S^q \to A \wedge A^\perp$. Clearly, "to be dual" is a symmetric relation. Furthermore, the space $\{A^\perp\}$ is unique up to stable homotopy equivalence. (Indeed, if there is another q-dual space A' then we have a natural in E isomorphism $\{A^\perp, E\} = \{A', E\}$.) In particular, the spaces $(A^\perp)^\perp$ and A are stable homotopy equivalent.

(c) Let $u : S^q \to A \wedge A^\perp$ and $v : S^q \to B \wedge B^\perp$ be two dualities, and let $f : A \to B$ be a map. Consider the isomorphism

$$D : \{A, B\} \xrightarrow{u_B} \{S^q, B \wedge A^\perp\} \xrightarrow{(v^{A^\perp})^{-1}} \{B^\perp, A^\perp\}$$

and a map $f^\perp : B^\perp \to A^\perp$ such that $D\{f\} = \{f^\perp\}$. So, $\{f^\perp\}$ is defined unique up to stable homotopy.

(d) Given A and B as in (c), consider a q-duality $S^q \to X \wedge X^\perp$. Then every map $g : B \to X$ yields a commutative diagram

$$
\begin{array}{ccc}
\{A, B\} & \xrightarrow{\;g_*\;} & \{A, X\} \\
{\scriptstyle D}\big\downarrow & & \big\downarrow{\scriptstyle D} \\
\{B^\perp, A^\perp\} & \xrightarrow{\;(g^\perp)^*\;} & \{X^\perp, A^\perp\}.
\end{array}
$$

2.1.5 Remark. Let X be a finite CW subspace of $\mathbb{R}^n$, and let U be a regular neighborhood of X. Then X is $(n-1)$-dual to $\mathbb{R}^n \setminus U$, Spanier–Whitehead [SW]. So, for every CW space X there exists N large enough and a space $X^\perp$ such that $X^\perp$ is N-dual to X.

2.1.6 Proposition. *Let $u : S^q \to A \wedge A^\perp$ be a q-duality between two finite CW-spaces. Then, for all i and π, the map u yields an isomorphism*

$$H_i(u; \pi) : \tilde{H}^i(A^\perp; \pi) \to \tilde{H}_{q-i}(A; \pi).$$

Proof. Recall that

$$H^n(A^\perp; \pi) \cong [A, K(\pi, n)]^\bullet \cong [S^N A, K(\pi, N + n)]^\bullet$$

where $K(\pi, i)$ is the Eilenberg–MacLane space. Because of Theorem 2.1.1, the last group coincides with $\{S^N A, K(\pi, N + n)\}$ for N large enough, and therefore

$$H^n(A^\perp; \pi) \cong \{S^N A, K(\pi, N + n)\} \text{ for } N \text{ large enough.}$$

Let $\varepsilon_n : SK(\pi, n) \to K(\pi, n + 1)$ be the adjoint map to the standard homotopy equivalence $K(\pi, n) \to \Omega K(\pi, n + 1)$, see e.g. [Sw]. G. Whitehead [Wh1] noticed an isomorphism

$$\tilde{H}_n(A; \pi) \cong \varinjlim [S^{N+n}, K(\pi, N) \wedge A]^\bullet.$$

Here $\varinjlim$ is the direct limit of the sequence

$$[S^{N+n}, K(\pi, N) \wedge A]^\bullet \longrightarrow [S^{N+n+1}, SK(\pi, N) \wedge A]^\bullet$$
$$\xrightarrow{\varepsilon_*} [S^{N+n+1}, K(\pi, N) \wedge A]^\bullet$$

(see [Gray, Ch. 18] or [Rud, II.3.24] for greater details). Since ε_n is an n-equivalence, and because of Theorem 2.1.1, we conclude that

$$\tilde{H}_n(A; \pi) \cong [S^{N+n}, K(\pi, N) \wedge A] \text{ for } N \text{ large enough.}$$

So, again because of Theorem 2.1.1,

$$\tilde{H}_n(A; \pi) \cong \{S^{N+n}, K(\pi, N) \wedge A\}$$

for N large enough.

Now, consider a q-duality $u : S^d \to A \wedge A^\perp$. Fix i and choose N large enough such that

$$\tilde{H}^i(A^\perp; \pi) \cong \{S^N A^\perp, K(\pi, N + i)\},$$
$$\tilde{H}_{q-i}(A; \pi) \cong \{S^{N+q}, K(\pi, N + i) \wedge A\}.$$

Put $K = K(\pi, N + i)$. By suspending the domain and the range, we get a duality (denoted also by u)

$$u : S^{N+d} \to A \wedge S^N A^\perp.$$

This duality yields the isomorphism

$$u^K : \tilde{H}^i(A^\perp; \pi) \cong \{S^N A^\perp, K\} \to \{S^{N+q}, K \wedge A\} = \tilde{H}_{d-i}(A; \pi),$$

and we set $H_i(u; \pi) := u^K$. $\square$

2.1.7 Remark. In fact, Proposition 2.1.6 gives us a criterion of whether u is a duality map. Moreover, Spanier–Whitehead [SW] *defined* the duality by means of 2.1.6. Cf. also [DP].

2.1.8 Definition. We dualize 1.5.5 and say that a pointed map $a : A \to S^k$ (or its stable homotopy class $a \in \{A, S^k\}$) is a *coreducibility* if the induced map

$$a^* : \widetilde{H}^i(S^k) \to \widetilde{H}^i(A)$$

is an isomorphism for $i \leqslant k$.

2.1.9 Proposition. *Let* $u : S^q \to A \wedge A^\perp$ *be a q-duality between two finite CW-spaces, and let $k \ll q$. A class $\alpha \in \{A^\perp, S^k\}$ is a coreducibility if and only if the class $\beta := u^{S^k}\alpha \in \{S^{q-k}, A\}$ is a reducibility.*

Proof. Let $H_i(u) : \widetilde{H}^i(A^\perp) \to \widetilde{H}_{d-i}(A)$ be the isomorphism described in 2.1.6. Note that the standard homeomorphism $v : S^q \to S^k \wedge S^{q-k}$ is a q-duality. It is easy to see that the diagram

$$
\begin{array}{ccc}
\widetilde{H}^i(A^\perp) & \xrightarrow{H_i(u)} & \widetilde{H}_{q-i}(A) \\
{\scriptstyle \alpha^*}\uparrow & & \uparrow{\scriptstyle \beta_*} \\
\widetilde{H}^i(S^k) & \xrightarrow{H_i(v)} & \widetilde{H}_{q-i}(S^{q-k})
\end{array}
$$

commutes. In particular, the left vertical arrow is an isomorphism if and only if the right one is. $\square$

2.1.10 Definition. Given a spherical fibration $\xi = \{p : E \to X\}$ over a finite CW-space B, an *automorphism* $\varphi : \xi \to \xi$ is a morphism over B

$$
\begin{array}{ccc}
E & \xrightarrow{g} & E \\
{\scriptstyle p}\downarrow & & \downarrow{\scriptstyle p} \\
B & \xrightarrow{1_B} & B
\end{array}
$$

where all maps of fibers g_b as in 1.3.1 are homotopy equivalences. We denote the set of fiber homotopy classes of automorphisms by $\mathrm{aut}\,\xi$.

Let $\sigma^k = \sigma_X^k$ denote the trivial S^{k-1}-bundle over X. Recall that G_n denotes the monoid of homotopy self-equivalences $S^{n-1} \to S^{n-1}$ equipped with the compact-open topology.

2.1.11 Proposition. *Let X be a connected CW space. There is a natural bijection*

$$\text{aut}\,\sigma^k = [X, G_k].$$

Proof. See [Br2, Prop. I.4.7]. $\qquad\qquad\qquad\qquad\qquad\qquad\qquad\qquad\square$

It will be convenient for us to interpret $\text{aut}\,\sigma^n_X$ in terms of automorphisms of the trivial vector bundle θ^n_X. We describe $\mathbb{R}^n$ as

$$\mathbb{R}^n = S^{n-1} \times [0, \infty)/\sim$$

where $(a, 0) \sim (b, 0)$ for all $a, b \in S^{n-1}$. Now, given an automorphism $\varphi : X \times S^{n-1} \to X \times S^{n-1}$ of σ^n, we define $\psi : X \times \mathbb{R}^n \to X \times \mathbb{R}^n$ by setting $\psi(x, r) = r\varphi(x, r)$. Clearly, ψ is completely determined by φ, and vice versa.

2.2 Thom Spaces. Proof of Theorem 1.5.6

Consider a closed connected n-dimensional PL manifold M and embed it in $\mathbb{R}^{N+n+k}$ with N large enough. Let $d : M \to M \times M$ be the diagonal map and $\theta^k = \theta^k_M$. Then $d^*(\nu^N \times \theta^k) = \nu^{N+k}$, and we have the adjoint map $\mathfrak{J} : \nu^{N+k} \to \nu^N \times \theta^k$. The map $\mathfrak{J}$ induces the map of Thom spaces

$$T\mathfrak{J} : T\nu^{N+k} \to T\nu^N \wedge T\theta^k.$$

Let $\iota : S^{N+n+k} \to T\nu^{N+k}$ be a collapsing map as in 1.5.3.

2.2.1 Theorem. *The map*

$$u : S^{N+n+k} \xrightarrow{\quad\iota\quad} T\nu^{N+k} \xrightarrow{\quad T\mathfrak{J}\quad} T\nu^N \wedge T\theta^k$$

is an $(N + n + k)$-duality map.

Proof. See Wall [W2, Theorem 3.2]. Cf. also [Atiyah]. $\qquad\qquad\qquad\square$

Milnor–Spanier [MiS] proved the duality between $T\nu$ and M^+, while the theorem tells more: it gives us an explicit form of the duality map. Note also that Switzer [Sw] proved Theorem 2.2.1 for orientable manifolds. Wall [W2] used cohomology with local coefficients in order to get rid of orientability restriction.

Recall that $T\theta^k = (M \times S^k)/M = S^k(M^+)$. Consider a surjective map $e : M^+ \to S^0$ and put $\varepsilon := S^k e : T\theta^k \to S^k$.

2.2.2 Proposition. *The collapsing map*

$$\iota : S^{N+n+k} \to T\nu^{N+k}$$

is $(N+n+k)$-dual to the map ε.

Proof. Consider two $(N+n+k)$-dualities

$$u : S^{N+n+k} \to T\nu^N \wedge T\theta^k$$

and

$$v : S^{N+n+k} \xrightarrow{=} S^{N+n} \wedge S^k.$$

We put $A = S^{N+n}$, $A^\perp = S^k$, $B = T\nu^N$, and $B^\perp = T\theta^k$. Then the duality isomorphism 2.1.4(c) gets the form (where $T\nu = T\nu^N$)

$$\{S^{N+n}, T\nu\} \xrightarrow{u_{T\nu}} \{S^{N+n+k}, T\nu \wedge S^k\} \xleftarrow{v^{S^k}} \{T\theta^k, S^k\}.$$

Now

$$u_{T\nu}(\iota) = \iota \wedge 1_{S^k} : S^{N+n+k} = S^{N+n} \wedge S^k \longrightarrow T\nu \wedge S^k, \text{ and}$$

$$v^{S^k}(\varepsilon) = 1_{T\nu} \wedge \varepsilon : S^{N+n+k} \longrightarrow T\nu \wedge T\theta^k \longrightarrow T\nu \wedge S^k.$$

So, $u_{T\nu}(\iota) = v^{S^k}(\varepsilon)$, and thus ι is dual to ε.

2.2.3 Corollary. *The collapsing map $\iota : S^{N+n} \to T\nu^N$ is a reducibility.*

Proof. Clearly, ε is a coreducibility because $e : \tilde{H}^0(M^+) \to \tilde{H}^0(S^0)$ is an isomorphism. Thus, the result follows from 2.1.9. □

Every automorphism $\psi : \theta^k \to \theta^k$ yields the automorphism $\psi \oplus 1_{\theta^k} : \theta^{2k} \to \theta^{2k}$. So, we get a homeomorphism of Thom spaces

$$T(\psi \oplus 1_{\theta^k}) : T\theta^{2k} \longrightarrow T\theta^{2k}.$$

Similarly, every automorphism $\psi : \theta^k \to \theta^k$ yields the authomorphiam

$$1_{\nu^{N+k}} \oplus \psi : \nu^{N+k} \oplus \theta^k \to \nu^{N+k} \oplus \theta^k.$$

So, we get a homeomorphism of Thom spaces

$$T(1_\nu \oplus \psi) : T\nu^{N+k} \longrightarrow T\nu^{N+k}.$$

The duality map from Theorem 2.2.1 yields an isomorphism

$$u := u^{S^k} : \{T\theta^k, S^k\} \to \{S^{N+n+k}, S^k \wedge T\nu^N\} = \{S^{N+n}, T\nu^N\}.$$

2.2.4 Proposition. *Let ψ be an automorphism of θ^k. The following diagram commutes.*

$$
\begin{array}{ccc}
\{T\theta^{2k}, S^{2k}\} & \xrightarrow{\;T(\psi \oplus 1_{\theta^k})_*\;} & \{T\theta^{2k}, S^{2k}\} \\
u\downarrow & & \downarrow u \\
\{S^{N+n+k}, T\nu^{N+k}\} & \xrightarrow{\;T(1_\nu \oplus \psi)_*\;} & \{S^{N+n+k}, T\nu^{N+k}\}
\end{array}
$$

Proof. This follows because $T(\psi \oplus 1_{\theta^k})$ is dual to $T(1_\nu \oplus \psi)$, see [Br2, Theorem I.4.16].   □

So, we have the $\operatorname{aut}\theta^k$-action

$$a_\nu : \operatorname{aut}\theta^k \times \{S^{N+n+k}, T\nu^{N+k}\} \to \{S^{N+n+k}, T\nu^{N+k}\},$$

$$a_\nu(\psi, \alpha) = T(1_\nu \oplus \psi)_*(\alpha).$$

Similarly, we have the action

$$a_s : \operatorname{aut}\theta^k \times \{T\theta^k, S^k\} \to \{T\theta^k, S^k\}, \quad a_s(\psi, \alpha) = T(\psi \oplus 1_{\theta^k}) * (\alpha).$$

2.2.5 Theorem. *The diagram*

$$
\begin{array}{ccc}
\operatorname{aut}\theta^k \times \{T\theta^k, S^{2k}\} & \xrightarrow{\;a_s\;} & \{T\theta^k, S^{2k}\} \\
1\times u\downarrow & & \downarrow u \\
\operatorname{aut}\theta^k \times \{S^{N+n+k}, T\nu^{N+k}\} & \xrightarrow{\;a_\nu\;} & \{S^{N+n+k}, T\nu^{N+k}\}
\end{array}
$$

commutes.

Proof. This follows from Proposition 2.2.4 and the definition of the mappings $D, a_\nu,$ and a_s.   □

Take k large enough and $N \geqslant k$ such that

$$\{T\theta^k, S^k\} = \pi^k(T\theta^k) \text{ and } \{S^{N+n}, T\nu^N\} = \pi_{N+n}(T\nu^N).$$

This is possible because of Theorem 2.1.1.

Let $\mathcal{R} \in \pi_{N+n}(T\nu^N)$ be the set of reducibilities, and let $\mathcal{C} \in \pi^k(T\theta^k)$ be the set of coreducibilities. Then, clearly, $a_\nu(\mathcal{R}) \subset \mathcal{R}$ and $a_s(\mathcal{C}) \subset \mathcal{C}$. (Here we write $\pi_{N+n}(T\nu^N)$ while we consider $\pi_{N+n+k}(T\nu^{N+k})$ in Proposition 2.2.4, but we do not care on this since N is large.) Therefore, in view of Proposition 2.1.9, the diagram from Theorem 1.2.1 yields the commutative diagram

$$
\begin{array}{ccc}
\operatorname{aut}\theta^k \times \mathcal{C} & \xrightarrow{\;a_\sigma\;} & \mathcal{C} \\
1\times D\downarrow & & \downarrow D \\
\operatorname{aut}\theta^k \times \mathcal{R} & \xrightarrow{\;a_\nu\;} & \mathcal{R}
\end{array}
\qquad (2.2.1)
$$

2.2.6 Theorem. *For every* $\alpha, \beta \in \mathcal{C}$ *there exists an automorphism* ψ *of* θ^k *such that* $a_s(\psi, \alpha) = \beta$. *Moreover, the map* $\psi : S^{k-1} \to S^{k-1}$ *is unique up to fiberwise homotopy. In other words, the action* $a_s : \operatorname{aut} \theta^k \times \mathcal{C} \to \mathcal{C}$ *is free and transitive.*

Proof. Recall that $T\theta^k = (M \times S^k)/M$. So, for every $m \in M$, a map $f : T\theta^k \to S^k$ yields a map $f_m : S^k_m \to S^k$ where S^k_m is the fiber over m. Furthermore, f represents a coreducibility if and only if all maps f_m belong to G_k. In other words, every coreducibility $T\theta^k \to S^k$ yields a homotopy class $M \to G_k$, and in fact we have a bijection $\mathcal{C} \to [M, G_k]$ there. Moreover, it is easy to see that, in view of Proposition 2.1.11, the action a_s coincides with the map

$$[M, G_k] \times [M, G_k] \to [M, G_k]$$

that is induced by the product in G_k, and the result follows.   □

2.2.7 Corollary. *The action* $a_\nu : \operatorname{aut} \theta^k \times \mathcal{R} \to \mathcal{R}$ *is free and transitive.*

Proof. This follows from Theorem 2.2.6 because D is an isomorphism.   □

Now we complete the proof of Theorem 1.5.6. Recall notation α, ν, η, and ι from Theorem 1.5.6: Here η is an arbitrary S^{N-1}-fibration such that $T\alpha$ is reducible and $\alpha : S^{N+k} \to T\eta$ is a reducibility for $T\eta$. Furthermore, ν is a normal bundle of M in $\mathbb{R}^{N+n}$ and $\iota : S^{N+n} \to T\nu$ is the collapsing map. We need to prove that there exists a fiberwise S^{N-1}-equivalence $\mu : \nu^\triangledown \to \eta$ with $(T\mu) \circ \iota \simeq \alpha$, and that such μ is unique up to fiberwise homotopy.

Choose any such S^{N-1}-equivalence $\psi : \eta \to \nu^\triangledown$ and consider the induced homotopy equivalence $T\psi : T\eta \to T\nu^\triangledown$. Clearly, the composition

$$\beta : S^{N+n} \xrightarrow{\ \alpha\ } T\eta \xrightarrow{\ T\psi\ } T\nu$$

is a reducibility. Furthermore, ι is also a reducibility. So, by 2.2.7, there exists an S^{N+k}-equivalence (automorphism) $\lambda : \nu^\triangledown \to \nu^\triangledown$ over M with $(T\lambda) \circ \beta = \iota$. Now, we define $\mu : \nu^\triangledown \to \eta$ to be the fiber homotopy inverse to $\lambda\psi$. (The existence of an inverse equivalence is a theorem of Dold [Dold], cf. also [May]). Clearly, $(T\mu) \circ \iota = \alpha$. This proves the existence of the required equivalence μ.

Finally, if there exists another equivalence $\mu' : \nu^\triangledown \to \eta$, then

$$(T(\mu^{-1} \circ \mu')) \circ \iota \simeq \iota : S^{N+n} \to T\nu^N.$$

Because of Corollary 2.2.7, the map $\mu^{-1} \circ \mu'$ is fiberwise homotopic to the identity morphism $1_{\nu^\triangledown}$, and so μ and μ' are homotopic over M. This proves the uniqueness of μ. Thus, Theorem 1.5.6 is proved.   □

2.3 Normal Morphisms, Normal Bordisms, and G/PL

Throughout the section we fix a closed orientable n-dimensional PL manifold M. Browder [Br2, Chapter II, §1]] introduced the following notion.

2.3.1 Definition. A *normal morphism at M* is a PL $\mathbb{R}^N$-morphism $\varphi : \nu_V \to \xi$ where $\xi = \{E(\xi) \to M\}$ is a PL $\mathbb{R}^N$-bundle over M, V is a closed PL manifold, $\nu_V = \{E(\nu) \to V\}$ is a normal PL $\mathbb{R}^N$-bundle of V in $\mathbb{R}^{N+n}$, and the map $f = \mathrm{bs}\,\varphi : V \to M$ is of degree 1. See the diagram

$$
\begin{array}{ccc}
E(\nu) & \xrightarrow{\ g\ } & E(\xi) \\
\downarrow & & \downarrow \\
V & \xrightarrow{\ f\ } & M.
\end{array}
$$

We denote the set of all normal morphisms at M by $\mathrm{Nor}(M)$. (For persons who ask whether $\mathrm{Nor}(M)$ is a set, we take the space $\mathbb{R}^\infty$ and assume that all the spaces in the Definition are contained in $\mathbb{R}^\infty$.)

2.3.2 Example. Let $h : V \to M$ be a homotopy equivalence and $g : M \to V$ a homotopy inverse map to h. Consider the normal bundle ν of V and put $\xi := g^*(\nu)$. Then $h^*(\xi) = h^*g^*\nu = \nu$. The adjoint morphism $\nu = h^*(\xi) \to \xi$ is a normal morphism.

It is worthy to note the following fact.

2.3.3 Proposition. *If $\varphi : \nu_V \to \xi$ is a normal morphism as in 2.3.1 then*

$$
S^{N+n} \xrightarrow{\ \mathrm{collapse}\ } T\nu_V \xrightarrow{\ T\varphi\ } T\xi
$$

is a reducibility.

Proof. Since $H_i(T\nu_V) = 0 = H_i(T\xi)$ for $i > N + n$, it suffices to prove that

$$
T\varphi : \mathbb{Z} = H_{N+n}(T\nu_V) \to H_{N+n}(T\xi) = \mathbb{Z}
$$

is an isomorphism. Because of functoriality of the Thom isomorphism (see e.g., [Rud, V.1.2]), we have the commutative diagram

$$
\begin{array}{ccc}
H_{N+n}(T\nu_V) & \xrightarrow{\ T\varphi\ } & H_{N+n}(T\xi) \\
\phi_1 \uparrow & & \uparrow \phi_2 \\
H_n(V) & \xrightarrow{\ f_*\ } & H_n(M)
\end{array}
$$

where ϕ_1, ϕ_2 are the Thom isomorphisms. Note that f_* is an isomorphism because $\deg f = 1$, and the result follows. $\qquad\square$

2.3.4 Construction–Definition. Let $\varphi : \nu_V \to \xi$ be a normal morphism at M and assume that $\dim \nu_V$ is large. Consider a collapsing map (homotopy class) $\iota : S^{N+n} \to T\nu_M$ as in 1.5.3. Since the map

$$\alpha : S^{N+n} \xrightarrow{\text{collapse}} T\nu_V \xrightarrow{T\varphi} T\xi$$

is a reducibility, there exists, by Theorem 1.5.6, a unique morphism of S^{N-1}-fibrations $\mu : \nu_M^\nabla \to \xi^\nabla$ with $\mu_*(\iota) = \alpha$. Now, the morphism

$$\nu_M^\nabla \xrightarrow{\mu_*} \xi^\nabla \xrightarrow{\text{classif}} (\gamma_{PL}^N)^\nabla$$

is a homotopy PL structure on ν_M. Thus, in view of 1.4.12, we get a homotopy class in $[M, G/PL]$. We denote by $f_\varphi : M \to G/PL$ any representative of this class.

2.3.5 Definition. The function

$$\Gamma = \Gamma_M : \text{Nor}(M) \longrightarrow [M, G/PL], \quad \varphi \mapsto [f_\varphi].$$

is called the *normal invariant* for M.

Probably, the reader noticed that we already defined normal invariant j_G in Definition 1.5.7. But there is no ambiguity. Indeed, the map

$$\{\text{homotopy equivalences } ? \to M\} \xrightarrow{2.3.2} \text{Nor } M \xrightarrow{\Gamma} [M, G/PL]$$

coincides with j_G.

2.3.6 Definition (Browder [Br2]). (a) A *normal bordism* between two normal morphisms $\varphi_i : \nu_{V_i} \to \xi$, $i = 0, 1$ at M is a PL $\mathbb{R}^N$-morphism $\Phi : \nu_W \to \xi$ where W is a compact PL manifold with $\partial W = V_0 \sqcup V_1$ and $\Phi|_{V_i} = \varphi_i, i = 0, 1$. Furthermore, $\nu_W = \{D \to W\}$ is the PL normal $\mathbb{R}^N$-bundle of W. So, Φ is a commutative diagram

$$
\begin{array}{ccc}
D & \longrightarrow & E \\
\downarrow & & \downarrow \\
W & \longrightarrow & M
\end{array}
$$

where $\xi = \{E \to M\}$ is the same bundle as in Definition 2.3.1.

(b) We say that two normal morphisms are *normally bordant* if there exists a normal bordism between these two normal morphisms. Clearly, "to be normally bordant" is an equivalence relation. The correspondence equivalence class is called a normally bordantness class or *normal bordism class*.

We would like to proclaim that Γ yields a bijection between the set of normally bordantness classes at M and the set $[M, G/PL]$. However, this is wrong. For example, if you compose the normal morphism φ as in Definition 2.3.6 with a PL isomorphism of bundles $\xi \to \xi'$ then you have the same element of $[M, G/PL]$, but, probably, a different normal bordism class, cf. [Br2, Lemma II.4.8]. Keeping this in mind, we have the following definition.

2.3.7 Definition. A *weak normal bordism* between two normal morphisms $\varphi_i : \nu_{V_i} \to \xi_i$, $i = 0, 1$ at M is a PL $\mathbb{R}^N$-morphism $\Psi : \nu_W \to \xi \times I$ where W is a compact PL manifold with $\partial W = V_0 \sqcup V_1$ and $\Psi|_{V_i} = \xi_i, i = 0, 1$. Moreover, $\xi_i = \xi|_{X \times \{i\}}, i = 0, 1$. Finally, $\Psi|_{V_0} = \varphi_0$ but

$$\Psi|_{V_1} = \{b\varphi_1 : \nu_1 \to \xi_1 \to \xi_0\}$$

where $b : \xi_1 \to \xi_0$ is a bundle isomorphism. Furthermore, $\nu_W = \{D \to W\}$ is the PL normal $\mathbb{R}^N$-bundle of W. See the diagram

$$
\begin{array}{ccc}
D & \longrightarrow & E \times I \\
\downarrow & & \downarrow \\
W & \longrightarrow & M \times I.
\end{array}
$$

2.3.8 Remark. Note that the word "bordism" is now used in place of the original term "cobordism" in [Br2, MM]. Browder [Br2] defined normal (co)bordism as in Definition 2.3.6. Madsen–Milgram [MM] used the term "normal (co)bordism" for what you called "weak normal bordism".

Similarly to Definition 2.3.6, "to be weakly normally bordant" is an equivalence relation. The equivalence classes are called the *weak normal bordism classes*. We denote by $[\varphi]$ the weak normal bordism class of φ and by $[\mathrm{Nor}\, M]$ the set of weak normal bordism classes at M.

2.3.9 Theorem. *If φ_0, φ_1 are two weakly normally bordant normal morphisms at M then*

$$f_{\varphi_0} \simeq f_{\varphi_1} : \mathrm{Nor}\, M \to [M, G/PL].$$

So, the map Γ yields a map

$$\widetilde{\Gamma} : [\mathrm{Nor}\, M] \to [M, G/PL], \quad \widetilde{\Gamma}[\varphi] = [\Gamma(\varphi)].$$

Moreover, the map $\widetilde{\Gamma}$ is a bijection.

Proof (sketch). This is a version of the Pontryagin–Thom theorem. We give an outline of the proof and leave the detail to the reader. Let Ψ be a weak normal bordism as in Definition 2.3.7. Follow 2.3.4 and construct a map $F_\Psi : M \times I \to G/PL$. Then $F_\Psi|_{M \times \{i\}} = f_{\varphi_i}$ for $i = 0, 1$. So, the above-mentioned map $\widetilde{\Gamma}$ is well-defined.

We construct the map $\Delta : [M, G/PL] \to [\text{Nor } M]$ that is inverse to $\widetilde{\Gamma}$. Let $\nu = \nu^N$ denote the normal bundle of M with $N \gg m$. Recall that, because of (1.4.6) and (1.4.12), there is a bijection

$$\{\text{homotopy PL structure on } \nu^N\} \longrightarrow [M, G/PL]. \qquad (2.3.1)$$

Consider a map $f : M \to G/PL$ and take the morphism of spherical fibrations $\varphi : \nu^\triangledown \to (\gamma_{PL}^N)^\triangledown$ that corresponds to f with respect to (2.3.1). Set $\xi = (\text{bs}\,\varphi)^* \gamma_{PL}^N$. Then we get the adjoint morphism of PL $\mathbb{R}^N$-bundles $\mathfrak{J} : \xi \to \gamma_{PL}^N$ which, in turn, yields a morphism of spherical fbration $\xi^\triangledown \to (\gamma_{PL}^N)^\triangledown$. Let $\psi : \nu^\triangledown \to \xi^\triangledown$ be the correcting morphism (of S^{N-1}-fibrations) for φ. Note that $T\nu^\triangledown = T\nu$ and $T\xi^\triangledown = T\xi$ where T denote the Thom finctor.

Let $c \colon S^{N+n} \to T\nu$ be the collapsing map. Consider the map

$$h \colon S^{N+n} \xrightarrow{\ c\ } T\nu \xrightarrow{\ T\psi\ } T\xi \qquad (2.3.2)$$

and deform h to map $t : S^{N+n} \to T\xi$ that is transverse to $M \subset T\xi$. Set $V := t^{-1}(M)$ and $b := t|_V : V \to M$. So, we get the b-adjoint PL $\mathbb{R}^N$-morphism

$$\mathfrak{J}_b = \mathfrak{J}_b^f : b^*\xi \to \xi, \quad \text{bs}(\mathfrak{J}_b) = b : V \to M. \qquad (2.3.3)$$

Note that $b^*\xi$ is the normal bundle of V in S^{N+n}. Furthermore, we (can) choose the orientation of V such that $\deg b = 1$, cf. [Br2, II.2.13]. In other words, the morphism (2.3.3) is a normal morphism at M.

Put $\Delta[f]$ to be the weak normal bordism class of $\mathfrak{J}_b$. Check that this weak normal bordism class is well-defined and that $\widetilde{\Gamma}$ and Δ are inverse to each other. Cf. also [Br2, Lemma II.4.8]. $\qquad\qquad\square$

Recall that a closed manifold is called *almost parallelizable* if it becomes parallelizable after deleting of a point. Note that every almost parallelizable manifold is orientable (e.g., because its first Stiefel–Whitney class is equal to zero).

2.3.10 Proposition. *For every almost parallelizable PL manifold V^m there exists a normal morphism with a base $V \to S^m$.*

Proof. We regard $S^m = \{(x_1, \ldots, x_{m+1}) \mid \sum x_i^2 = 1\}$ as the union of two discs, $S^m = D_+ \cup D_-$, where

$$D_+ = \{x \in S^m | x_{m+1} \geqslant 0\}, \quad D_- = \{x \in S^m | x_{m+1} \leqslant 0\}.$$

Take a map $b : V \to M$ of degree 1. We can assume that there is a small closed disk D_0 in V such that $b_+ := b|_{D_0} : D_0 \to D_+$ is a PL homeomorphism. We set $W = V \setminus (\operatorname{Int} D_0)$. Since W is parallelizable, there exists a PL morphism $\varphi_- : \nu_V|_W \to \theta_{D_-}$ such that $b|_W : W \to D_-$ is the base of φ_-. Furthermore, since b_+ is a PL homeomorphism, there exists a morphism $\varphi_+ : \nu_V|_{D_0} \to \theta_{D_+}$ over b_+ such that φ_+ and φ_- coincide over $b|_{\partial W} : \partial W \to S^{m-1}$. Together φ_+ and φ_- give us a PL morphism $\varphi : \nu_V \to \xi$ where ξ is a PL bundle over S^m. Clearly, φ is a normal morphism with the base b. □

2.4 The Sullivan Map $s : [M, G/PL] \to P_{\dim M}$

We define the groups P_i by setting

$$P_i = \begin{cases} \mathbb{Z} & \text{if } i = 4k, \\ \mathbb{Z}/2 & \text{if } i = 4k - 2, \\ 0 & \text{if } i = 2k - 1 \end{cases}$$

where $k \in \mathbb{N}$.

2.4.1 Definition. Given a closed connected orientable n-dimensional PL manifold M, we define a map

$$s : [M, G/PL] \to P_n$$

as follows. Given a homotopy class $f : M \to G/PL$, consider a normal morphism

$$\mathfrak{I} = \mathfrak{I}_b : b^*\xi \to \xi, \quad \operatorname{bs}(\mathfrak{I}) = b : V \to M$$

such that $\widetilde{\Gamma}[\mathfrak{I}] = [f]$, see Theorem 2.3.9, especially (2.3.3).

For $n = 4k$, let ψ be the symmetric bilinear intersection form on

$$\operatorname{Ker}\{b_* : H_{2k}(V; \mathbb{Q}) \to H_{2k}(M; \mathbb{Q})\}.$$

We define $s(f) = \sigma(\psi)/8$ where $\sigma(\psi)$ is the signature of ψ. It is well known that $\sigma(\psi)$ is divisible by 8 by purely algebraic reasons, (see e.g. [Br2, Proposition III.1.4]), and so $s(f) \in \mathbb{Z}$.

We use the notation as in (2.3.3) and let $\sigma(V), \sigma(M)$ be the signature of the manifold V, M, respectively. It is clear that $\sigma(\psi) = \sigma(V) - \sigma(M)$, and so

$$s(f) = \frac{\sigma(V) - \sigma(M)}{8}.$$

For $n = 4k - 2$, we define $s(f)$ to be the Kervaire invariant of the normal morphism $\mathfrak{J}_b$, see e.g. [Br2].

The routine Pontryagin–Thom arguments show that s is well-defined, i.e., it does not depend on the choice of the map f in the homotopy class in $[M, G/PL]$ and of the normal morphism $\mathfrak{J}_b$. See [Br2, Ch. III, §4] or [N1] for details.

In particular, if b is a homotopy equivalence then $s(f) = 0$.

2.4.2 Theorem (Sullivan [Sul2]). *For any closed simply-connected PL manifold M of dimension $\geqslant 5$, the sequence*

$$0 \longrightarrow \mathcal{S}_{PL}(M) \xrightarrow{\ j_G\ } [M, G/PL] \xrightarrow{\ s\ } P_{\dim M}$$

is exact, i.e., j_G is injective and $\operatorname{Im} j_G = s^{-1}(0)$.

The above-mentione exact sequence is called the *surgery exact sequence*.

Proof. You can find the proof of the smooth analogue of the theorem in [Br2, II.4.10 and II.4.11]. Let me note the changes that serve the PL case. Literally, Browder [Br2, II.4.10] considered the exact sequence

$$P_{m+1} \xrightarrow{\ \omega\ } \mathcal{S}(M) \longrightarrow [M, G/O] \xrightarrow{\ s\ } P_m$$

where M runs over smooth m-manifolds and $\mathcal{S}(M)$ is the set of concordance classes of smooth structures. Furthermore, the map (action) $\omega : P_{m+1} \to \mathcal{S}(M)$ appears as an action of a certain subgroup of the group of homotopy spheres Θ_m on $\mathcal{S}(M)$. So, the PL analog of ω is the zero map because, by Theorem 1.4.15, every homotopy sphere of dimension $\geqslant 5$ is PL homeomorphic to the standard sphere. $\square$

Given a map $f : M \to G/PL$, it is useful to introduce the notation $s(M, f) := s([f])$ where $[f]$ is the homotopy class of f.

Now we consider the important case when M is a sphere. Because of the additive property of signature and Kervaire invariant (see [Br2, II.1.4]), the map $s : [S^k, G/PL] \to P_k$ is a group homomorphism for all $k \in \mathbb{N}$.

2.4.3 Theorem. (i) *The homomorphism* $s : [S^{4i}, G/PL] \to \mathbb{Z}$ *is surjective for all* $i > 1$.

(ii) *The homomorphism* $s : [S^{4i-2}, G/PL] \to \mathbb{Z}/2$ *is surjective for all* $i > 0$.

(iii) *The image of the homomorphism* $s : [S^4, G/PL] \to \mathbb{Z}$ *is the subgroup of index* 2.

Proof. (i) For every $k > 1$ Milnor constructed a parallelizable $4k$-dimensional smooth manifold W^{4k} of signature 8 and such that ∂W is a homotopy sphere, see [Br2, V.2.9]. Since, by Theorem 1.4.15, every homotopy sphere of dimension $\geqslant 5$ is PL homeomorphic to the standard one, we can form a closed PL manifold

$$V^{4k} := W \cup_{S^{4k-1}} D^{4k}$$

of signature 8. Because of Proposition 2.3.10, there exists a normal morphism φ with the base $V^{4k} \to S^{4k}$. Because of Theorem 2.3.9, we have $\widetilde{\Gamma}[\varphi] = [f]$ for a certain map $f : S^{4k} \to G/PL$. Hence,

$$s(S^{4k}, f) = \frac{\sigma(V^{4k}) - \sigma(S^{4k})}{8} = 1.$$

If you take an m-fold boundary connected sum $W \# \cdots \# W$, you get a closed manifold

$$V_m = W \# \cdots \# W \cup_{S^{4k-1}} D^{4k}$$

and a map $f_m : S^{4k} \to G/PL$ with $s(S^{4k}, f_m) = m$. Thus, s is surjective.

(ii) The proof is similar to that of (i), but we must use (4k-2)-dimensional parallelizable Kervaire manifolds W, $\partial W = S^{4k-3}$ of the Kervaire invariant one, see [Br2, V.2.11].

(iii) The Kummer algebraic surface gives us an example of 4-dimensional almost parallelizable smooth manifold of the signature 16, see [K2]. So, $\operatorname{Im} s \supset 2\mathbb{Z}$.

Now suppose that there exists a map $f : S^4 \to G/PL$ with $s(S^4, f) = 1$. Then there exists a normal morphism with the base $V^4 \to S^4$ and such that V^4 has signature 8. Furthermore, since normal bundle of V^4 is induced from a bundle over S^4, we conclude that $w_1(V^4) = 0 = w_2(V^4)$. But this contradicts Rokhlin Theorem 1.8.1. $\qquad\square$

2.4.4 Corollary. *We have* $\pi_{4i}(G/PL) = \mathbb{Z}$, $\pi_{4i-2}(G/PL) = \mathbb{Z}/2$, *and* $\pi_{2i-1}(G/PL) = 0$ *for every* $i > 0$. *Moreover, the homomorphism*

$$s : [S^k, G/PL] \to P_k$$

is an isomorphism for $k \neq 4$, while for $k = 4$ it has the form

$$\mathbb{Z} = \pi_4(G/PL) \xrightarrow{\ s\ } P_4 = \mathbb{Z}, \quad a \mapsto 2a.$$

Proof. First, if $k > 4$ then, because of the Smale Theorem 1.4.15, $\mathcal{S}_{PL}(S^k)$ is the one-point set. Hence the map s is injective by Theorem 2.4.2, and thus it is an isomorphism by Theorem 2.4.3(i,ii).

If $k \leqslant 4$ then $\pi_k(PL/O) = 0$, cf. Remark 1.7.8. So, $\pi_k(G/PL) = \pi_k(G/O)$. Moreover, the forgetful map $\pi_k(BO) \to \pi_k(BG)$ coincides with the Whitehead J-homomorphism. So, we have the long exact sequence

$$\cdots \to \pi_k(G/O) \to \pi_k(BO) \xrightarrow{\ J\ } \pi_k(BG) \to \pi_{k-1}(G/O) \to \cdots.$$

For $k \leqslant 5$ all the groups $\pi_k(BO)$ and $\pi_k(BG)$ are known (note that $\pi_k(BG)$ is the stable homotopy group $\pi_{k+N-1}(S^N)$), and it is also known that J is an epimorphism for $k = 1, 2, 4, 5$, see [Ad]. Thus, $\pi_k(G/O) \cong P_i$ for $k \leqslant 4$.

The last assertion (about s in dimension 4) follows from Theorem 2.4.3(iii). $\qquad\qquad\qquad\qquad\qquad\qquad\qquad\qquad\qquad\qquad\qquad\square$

2.5 The Homotopy Type of $G/PL[2]$

Recall that, given a space X and an Abelian group π, we allow us to ignore the difference between elements of $H^n(X;\pi)$ and maps (homotopy classes) $X \to K(\pi, n)$.

2.5.1 Notation. Given a prime p, let $\mathbb{Z}[p]$ be the subring of $\mathbb{Q}$ consisting of all irreducible fractions with denominators relatively prime to p, and let $\mathbb{Z}[1/p]$ be the subgroup of $\mathbb{Q}$ consisting of the fractions $m/p^k, m \in \mathbb{Z}$. Given a simply-connected space X, we denote by $X[p]$ and $X[1/p]$ the $\mathbb{Z}[p]$- and $\mathbb{Z}[1/p]$-localization of X, respectively. Furthermore, we denote by $X[0]$ the $\mathbb{Q}$-localization of X. For the definitions, see [HMR].

2.5.2 Proposition (Sullivan [Sul1, Sul2]). *For every $i > 0$ there are cohomology classes*

$$K_{4i} \in H^{4i}(G/PL; \mathbb{Z}[2]), \quad K_{4i-2} \in H^{4i-2}(G/PL; \mathbb{Z}/2)$$

such that

$$s(M^{4i}, f) = \langle f^* K_{4i}, [M] \rangle$$

for every closed connected oriented PL manifold M, and

$$s(N^{4i-2}, f) = \langle f^* K_{4i-2}, [N]_2 \rangle$$

for every closed connected manifold N. Here $[M] \in H^{4i}(M)$ is the fundamental class of M, $[N]_2 \in H^{4i-2}(N; \mathbb{Z}/2)$ is the modulo 2 fundamental class of N, and $\langle -, - \rangle$ is the Kronecker pairing.

Proof. Let $MSO_*(-)$ denote the oriented bordism theory, see e.g., [Rud]. Recall that oriented bordant (oriented) manifolds have equal signatures, see e.g., [MS]. So, if two oriented singular manifolds $f : M^{4i} \to G/PL$ and $g : N^{4i} \to G/PL$ are bordant then $s(M, f) = s(N, g)$. Now, we define a homomorphism

$$\widetilde{s} : MSO_{4i}(G/PL) \to \mathbb{Z}, \quad \widetilde{s}[M, f] = s(M, f)$$

where $[M, f]$ is the bordism class of $f : M \to G/PL$.

It is well known that the Steenrod–Thom homomorphism

$$t : MSO_*(-) \otimes \mathbb{Z}[2] \to H_*(-; \mathbb{Z}[2])$$

splits, i.e., there is a natural homomorphism

$$v : H_*(-; \mathbb{Z}[2]) \to MSO_*(-) \otimes \mathbb{Z}[2]$$

such that $tv = 1$ (a theorem of Wall [W1], see also [Stong, Astey] and especially [Rud, Theorem IV.6.5]). In particular, we have a natural homomorphism

$$\widehat{s} : H_{4i}(G/PL; \mathbb{Z}[2]) \xrightarrow{\ v\ } MSO_{4i}(G/PL) \otimes \mathbb{Z}[2] \xrightarrow{\ \widetilde{s}\ } \mathbb{Z}.$$

Since the evaluation map

$$\mathrm{ev} : H^*(X; \mathbb{Z}[2]) \to \mathrm{Hom}(H_*(X; \mathbb{Z}[2]), \mathbb{Z}[2]),$$

$$(\mathrm{ev}(u))(v) = \langle u, v \rangle, \quad u \in H^*(X; \mathbb{Z}[2]), \ v \in H_*(X; \mathbb{Z}[2])$$

is surjective for all X, there exists a class $K_{4i} \in H^{4i}(G/PL; \mathbb{Z}[2])$ such that $\mathrm{ev}(K_{4i}) = \widehat{s}$. Now

$$s(M, f) = \widehat{s}(f_*[M]) = \langle K_{4i}, f_*[M] \rangle = \langle f^* K_{4i}, [M] \rangle.$$

So, we constructed the desired classes K_{4i}.

The construction of classes K_{4i-2} is similar. Let $MO_*(-)$ denoted the non-oriented bordism theory. Then the map s yields a homomorphism

$$\widetilde{s} : MO_{4i-2}(G/PL) \to \mathbb{Z}/2.$$

Furthermore, there exists a natural map $H_*(-; \mathbb{Z}/2) \to MO_*(-)$ which splits the modulo 2 Steenrod–Thom homomorphism, and so we have a homomorphism

$$\widehat{s} : H_{4i-2}(G/PL; \mathbb{Z}/2) \xrightarrow{\hspace{2cm}} MO_{4i-2}(G/PL) \otimes \mathbb{Z}[2] \xrightarrow{\ \widetilde{s}\ } \mathbb{Z}/2$$

with $\widehat{s}(f_*([M]_2)) = s(M, f)$. Now we can complete the proof similarly to the case of classes K_{4i}. $\qquad\square$

We set

$$\Pi := \prod_{i>1} \left(K(\mathbb{Z}[2], 4i) \times K(\mathbb{Z}/2, 4i-2) \right). \qquad (2.5.1)$$

Together the classes

$$K_{4i} : G/PL \to K(\mathbb{Z}[2], 4i), i > 1$$

and

$$K_{4i-2} : G/PL \to K(\mathbb{Z}/2, 4i-2), i > 1$$

yield a map

$$K : G/PL \to \Pi \qquad (2.5.2)$$

such that for each $i > 1$ the map

$$G/PL \xrightarrow{\ K\ } \Pi \xrightarrow{\ \text{projection}\ } K(\mathbb{Z}[2], 4i)$$

coincides with K_{4i} and the map

$$G/PL \xrightarrow{\ K\ } \Pi \xrightarrow{\ \text{projection}\ } K(\mathbb{Z}/2, 4i-2)$$

coincides with K_{4i-2}.

2.5.3 Lemma. *The map*

$$K[2] : G/PL[2] \to \Pi$$

induced an isomorphism of homotopy groups in dimensions ≥ 5.

Proof. This follows from Theorem 2.4.3 and Corollary 2.4.4.   □

Let Y be the Postnikov 4-stage of G/PL. So, we have a map

$$\psi : G/PL \longrightarrow Y \qquad (2.5.3)$$

that induces an isomorphism of homotopy groups in dimension $\leqslant 4$. Consider the map

$$\phi : G/PL[2] \to Y[2] \times \Pi, \quad \phi(x) = (\psi[2](x), K[2](x)).$$

2.5.4 Theorem. *The map*

$$\phi : G/PL[2] \to Y[2] \times \Pi$$

is a homotopy equivalence.

Proof. The maps

$$\phi_* : \pi_i(G/PL[2] \to \pi_i(Y[2] \times \Pi)$$

are isomorphisms for all i. Indeed, for $i \leqslant 4$ this holds since ψ is the Postnikov 4-approximation of G/PL, for $i \geqslant 5$ it follows from 2.5.3. Thus, ϕ is a homotopy equivalence by the Whitehead Theorem. $\qquad\square$

Now we discuss the space Y in greater detail. We have $\pi_2(Y) = \mathbb{Z}/2$, $\pi_4(Y) = \mathbb{Z}$, and $\pi_i(Y) = 0$ otherwise. So, we have a $K(\mathbb{Z}, 4)$-fibration

$$K(\mathbb{Z}, 4) \xrightarrow{\ i\ } Y \xrightarrow{\ p\ } K(\mathbb{Z}/2, 2) \qquad (2.5.4)$$

whose characteristic class is the Postnikov invariant

$$\kappa \in H^5(K(\mathbb{Z}/2, 2); \mathbb{Z})$$

of Y. We shall see in Theorem 2.5.8 below that

$$\kappa = \delta Sq^2 \iota_2 \neq 0.$$

Thus, κ is also the first non-zero Postnikov invariant of G/PL.

2.5.5 Lemma. There exists a map

$$g : \mathbb{CP}^2 \to G/PL$$

such that $s(\mathbb{CP}^2, g) = 1 = \langle K_4, g_*[\mathbb{CP}^2] \rangle$.

Proof. Let η denote the canonical complex line bundle over $\mathbb{CP}^2$. First, we prove that 24η is fiberwise homotopy trivial. Let $H : S^3 \to S^2$ be the Hopf map. Consider the Puppe sequence

$$S^3 \xrightarrow{\ H\ } S^2 \longrightarrow \mathbb{CP}^2 \longrightarrow S^4$$

and the induced exact sequence

$$[S^3, BG] \xleftarrow{\ H^*\ } [S^2, BG] \longleftarrow [\mathbb{CP}^2, BG] \longleftarrow [S^4, BG].$$

Let π_n^S denote the n-th stable homotopy group $\pi_{n+N}(S^N)$, N large. Note that $[S^n, BG] = \pi_{n-1}^S$. We have $[S^4, BG] = \pi_3^S = \mathbb{Z}/24$, [Hat]. Furthermore, the homomorphism

$$\mathbb{Z}/2 = [S^3, BG] \xleftarrow{\ H^*\ } [S^2, BG] = \mathbb{Z}/2$$

is an isomorphism, because the suspension $S^N H : S^{N+3} \to S^{N+2}$ is the generator of $\pi_{N+3} S^{N+2} = \pi_1^S = \mathbb{Z}/2$ for N large. Hence, $[\mathbb{CP}^2, BG]$ is a quotient group of $\mathbb{Z}/24$. In particular, 24η is fiberwise homotopy trivial.

Let p_1 and L_1 denote the first Pontryagin class and first Hirzebruch class, respectively. Recall that $L_1 = p_1/3$, see [MS]. Since $\langle p_1(\eta), [\mathbb{CP}^2]\rangle = 1$, we have $\langle p_1(24\eta), [\mathbb{CP}^2]\rangle = 24$, and therefore $\langle L_1(24\eta), [\mathbb{CP}^2]\rangle = 8$.

Let ν be the normal $\mathbb{R}^N$-bundle over $\mathbb{CP}^2$, N large, and let ξ be an $\mathbb{R}^{N-48}$-bundle over $\mathbb{CP}^2$ such that $\nu \cong \xi \oplus 24\eta$. Then ν^∇ and ξ^∇ are fiberwise homotopy equivalent. Take a fiberwise homotopy equivalence $u : \nu^\nabla \to \xi^\nabla$. Because of 2.3.1, u yields a map $g : \mathbb{CP}^2 \to G/PL$, and (in notation (2.3.3) and 2.4.1) we have

$$8s(\mathbb{CP}^2, g) = \sigma(V) - \sigma(\mathbb{CP}^2) = \langle L_1(\nu), [\mathbb{CP}^2]\rangle - \langle L_1(\nu_V), [V]\rangle$$
$$= \langle L_1(\nu), [\mathbb{CP}^2]\rangle - \langle L_1(\xi), [\mathbb{CP}^2]\rangle$$

because $b^*\xi$ is the normal bundle of V in S^{N+n}. Furthermore,

$$\langle L_1(\nu), [\mathbb{CP}^2]\rangle - \langle L_1(\xi), [\mathbb{CP}^2]\rangle = \langle L_1(\nu - \xi), [\mathbb{CP}^2]\rangle = \langle L_1(24\eta), [\mathbb{CP}^2]\rangle = 8.$$

Thus, $s(\mathbb{CP}^2, g) = 8/8 = 1$, and $\langle K_4, g_*[\mathbb{CP}^2]\rangle = 1$. $\square$

Let $h : \pi_4(G/PL) \to H_4(G/PL)$ be the Hurewicz homomorphism. Let tors denote the torsion subgroup of $H_4(G/PL)$.

2.5.6 Lemma. *The homomorphism*

$$a : \pi_4(G/PL) \xrightarrow{\ h\ } H_4(G/PL) \xrightarrow{\ \text{quotient}\ } H_4(G/PL)/\,\text{tors} = \mathbb{Z}$$
is not surjective.

Proof. Consider the Leray–Serre spectral sequence of the fibration (2.5.4) and note that $H_4(Y)/\,\text{tors} = \mathbb{Z}$, because $H_4(K(\mathbb{Z}, 4)) = \mathbb{Z}$ and all the groups $\widetilde{H}_i(K(\mathbb{Z}/2, 2))$ are finite. Furthermore, $H_4(G/PL) \cong H_4(Y)$ since Y is a Postnikov 4-stage of G/PL. Thus, $H_4(G/PL)/\,\text{tors} = \mathbb{Z}$.

Because of 2.4.4 and 2.5.2, the subgroup $\langle K^4, \text{Im}\, a \rangle$ of $\mathbb{Z}$ consists of even numbers. On the other hand, $\langle K_4, g_*[\mathbb{CP}^2]\rangle = 1$ by Lemma 2.5.5. Thus, the image of $g_*[\mathbb{CP}^2]$ in $H_4(G/PL)/\,\text{tors}$ does not belong to $\text{Im}\, a$. $\square$

Consider the short exact sequence

$$0 \longrightarrow \mathbb{Z} \xrightarrow{\ 2\ } \mathbb{Z}[2] \xrightarrow{\ \rho\ } \mathbb{Z}/2 \longrightarrow 0$$

where 2 over the arrow means multiplication by 2 and ρ is the modulo 2 reduction. This exact sequence yields the Bockstein exact sequence

$$\cdots \longrightarrow H^n(X; \mathbb{Z}) \xrightarrow{\ 2\ } H^n(X; \mathbb{Z}) \xrightarrow{\ \rho_*\ } H^n(X; \mathbb{Z}/2)$$
$$\xrightarrow{\ \delta\ } H^{n+1}(X; \mathbb{Z}) \longrightarrow \cdots . \tag{2.5.5}$$

Put $X = K(\mathbb{Z}/2, n)$ and consider the fundamental class

$$\iota_n \in H^n(K(\mathbb{Z}/2, n); \mathbb{Z}/2).$$

Then we have the class $\delta := \delta(\iota_n) \in H^{n+1}(K(\mathbb{Z}/2, n); \mathbb{Z})$. According to what we said above, we regard δ as a map $\delta : K(\mathbb{Z}/2, n) \to K(\mathbb{Z}, n+1)$ and/or the cohomology operation

$$\delta : H^n(-; \mathbb{Z}/2) \to H^{n+1}(-; \mathbb{Z}).$$

2.5.7 Lemma. *We have:* $H^{n+3}(K(\mathbb{Z}/2, n)) = \mathbb{Z}/2 = \delta Sq^2\iota_n$ *for all* $n \geqslant 4$.

Proof. Put $\iota = \iota_n$ and $K = K(\mathbb{Z}/2, n)$. Remark that all group $\widetilde{H}^i(K)$ are 2-primary in view of Serre Class Theory, [MT, Ch. 10]. Consider the Universal Coefficient Theorem

$$0 \longrightarrow H^{n+2}(K) \otimes \mathbb{Z}/2 \longrightarrow H^{n+2}(K; \mathbb{Z}/2)$$
$$\longrightarrow \mathrm{Tor}(H^{n+3}(K), \mathbb{Z}/2) \longrightarrow 0.$$

Note that $H^{n+2}(K; \mathbb{Z}/2) = \mathbb{Z}/2 = \{Sq^2\iota\}$ [MT, Ch. 9]. Furthermore, $Sq^3\iota \neq 0$, and so the group $H^{n+3}(K)$ contains a non-zero element $\delta Sq^2\iota$. Hence, $\mathrm{Tor}(H^{n+3}(K), \mathbb{Z}/2) = \mathbb{Z}/2$, and so the 2-primary group $H^{n+3}(K)$ is cyclic. Hence, the cyclic group $H^{n+3}(K)$ generates $\delta Sq^2\iota$ since

$$H^{n+3}(K; \mathbb{Z}/2) = \{Sq^3\iota, Sq^2 Sq^1 \iota\}$$

and $\rho_* \delta Sq^2\iota = Sq^3\iota$.

Finally, $\delta Sq^2\iota$ has the order 2 since $2\delta = 0$.   □

Recall that we denote the characteristic class of the fibration (2.5.4) by $\kappa \in H^5(K(\mathbb{Z}/2, 2))$.

2.5.8 Theorem. *We have* $\kappa = \delta Sq^2\iota_2$, *and it is an element of order 2 in the group* $H^5(K(\mathbb{Z}/2, 2)) = \mathbb{Z}/4$. *So,* κ *is the first non-trivial Postnikov invariant of* G/PL.

Proof. Note that $\kappa \neq 0$ because of 2.5.6. Indeed, otherwise $Y \simeq K(\mathbb{Z}, 4) \times K(\mathbb{Z}/2, 2)$. But this contradicts Lemma 2.5.6.

Let Ω be the loop functor on category of topological spaces and maps. Since G/PL is an infinite loop space, the Postnikov invariant κ of G/PL can be written as $\Omega^N a_N$ for all N and suitable

$$a_N \in [K(\mathbb{Z}/2, N+2), K(\mathbb{Z}.N+5)] = H^{N+5}(K(\mathbb{Z}/2, N+2); \mathbb{Z}).$$

By Lemma 2.5.7, the last group is equal to $\mathbb{Z}/2$ for $N > 5$. So, κ has the order 2. It is easy to see that

$$H^5(K(\mathbb{Z}/2, 2)) = \mathbb{Z}/4 = \{x\}$$

with $2x = \delta Sq^2\iota_2$, see e.g. [Rud, Lemma VI.2.7].

Thus, $\kappa = \delta Sq^2\iota_2$.   □

2.5.9 Lemma. *Let X be a finite CW-space such that the groups $H_i(X)$ are torsion free for all i. Let W be an infinite loop space such that groups $\pi_i(W)$ have no odd torsion for all i. Then the group $[X, W]$ is torsion free. In particular, the group $[X, G/PL[1/2]]$ is torsion free.*

Proof. It suffices to prove that $[X, W[p]]$ is torsion free for every odd prime p. Note that $W[p]$ is an infinite loop space since W is. So, there exists a connected p-local spectrum E such that

$$\widetilde{E}^0(-) = [-, W[p]] = [-, W \otimes \mathbb{Z}[p]].$$

Moreover, $E^{-i}(\mathrm{pt}) = \pi_i(E) = \pi_i(W) \otimes \mathbb{Z}[p]$. So, because of the isomorphism $\widetilde{E}^0(X) \cong [X, W[p]]$, it suffices to prove that $E^*(X)$ is torsion free. Consider the Atiyah–Hirzebruch spectral sequence for $E^*(X)$. Its initial term is torsion free because $E^*(\mathrm{pt})$ and $H^*(X)$ are torsion free. Hence, the spectral sequence degenerates, and thus the group $E^*(X)$ is torsion free.   □

2.5.10 Proposition. *Let X be a finite CW-space such that the group $H_*(X)$ is torsion free. Let $f : X \to G/PL$ be a map such that $f^* K_{4i} = 0$ and $f^* K_{4i-2} = 0$ for all $i \geqslant 1$. Then f is inessential.*

Proof. Consider the commutative square

$$
\begin{array}{ccc}
G/PL & \xrightarrow{\ l_1\ } & G/PL[2] \\
{\scriptstyle l_2}\downarrow & & \downarrow{\scriptstyle l_3} \\
G/PL[1/2] & \xrightarrow{\ l_4\ } & G/PL[0]
\end{array}
$$

where the horizontal maps are the $\mathbb{Z}[2]$-localizations and the vertical maps are the $\mathbb{Z}[1/2]$-localizations. Because of 2.5.4, $[X, G/PL]$ is a finitely generated Abelian group, and so it suffices to prove that both $l_1 \circ f$ and $l_2 \circ f$ are inessential. First, we remark that $l_2 \circ f$ is inessential whenever $l_1 \circ f$ is. Indeed, since $H_*(X)$ is torsion free, the group $[X, G/PL[1/2]]$ is torsion free by 2.5.9. Now, if $l_1 \circ f$ is inessential then $l_3 \circ l_1 \circ f$ is inessential, and hence $l_4 \circ l_2 \circ f$ is inessential. Thus, $l_2 \circ f$ is inessential since $[X, G/PL[1/2]]$ is torsion free.

So, it remains to prove that $l_1 \circ f$ is inessential.

Clearly, the equalities $f^* K_{4i} = 0$ and $f^* K_{4i-2} = 0$, $i > 0$, imply that the map

$$
X \longrightarrow G/PL \xrightarrow{\ l_1\ } G/PL[2] \simeq Y[2] \times \Pi \xrightarrow{\ p_2\ } \Pi
$$

is inessential. So, it remains to prove that the map

$$g : X \xrightarrow{\ f\ } G/PL \xrightarrow{\ l_1\ } G/PL[2] \simeq Y[2] \times \Pi \xrightarrow{\ p_1\ } Y[2]$$

is inessential.

It is easy to see that $H^4(Y[2]; \mathbb{Z}[2]) = \mathbb{Z}[2]$, see e.g., [Rud, VI.2.9(i)]. Let $u \in H^4(Y; \mathbb{Z}[2])$ be a free generator of the free $\mathbb{Z}[2]$-module $H^4(Y; \mathbb{Z}[2])$. The fibration (2.5.4) gives us the following diagram with the exact row:

$$H^4(X; \mathbb{Z}[2]) \xrightarrow{\ i_*\ } [X, Y[2]] \xrightarrow{\ p_*\ } H^2(X; \mathbb{Z}/2)$$
$$\downarrow{\scriptstyle u_*}$$
$$H^4(X; \mathbb{Z}[2])$$

Note that

$$u_* i_* : \mathbb{Z}[2] \to \mathbb{Z}[2]$$

is the multiplication by 2ε where ε is an invertible element of the ring $\mathbb{Z}[2]$, see e.g., [Rud, VI.2.9(ii)]. Since $f^* K_2 = 0$, we conclude that $p_*(g) = 0$, and so $g = i_*(a)$ for some $a \in H^4(X; \mathbb{Z}[2])$. Now,

$$0 = u_*(g) = u_*(i_* a) = 2a\varepsilon.$$

But $H^*(X; \mathbb{Z}[2])$ is torsion free, and thus $a = 0$. □

For completeness, we mention also that $G/PL[1/2] \simeq BO[1/2]$. So, there is a Cartesian square (see [MM, Sul2])

$$
\begin{array}{ccc}
G/PL & \longrightarrow & \Pi \times Y \\
\downarrow & & \downarrow \\
BO[1/2] & \xrightarrow{\ \text{ph}\ } & \prod K(\mathbb{Q}, 4i)
\end{array}
$$

where ph is the Pontryagin character.

2.6 Splitting Theorems

2.6.1 Definition. (a) Let A^{n+k} and W^{n+k} be two connected PL manifolds (without boundaries), and let M^n be a closed PL submanifold of A. We say that a map $g : W^{n+k} \to A^{n+k}$ *is splittable along* M^n if there exists a homotopy

$$g_t : W^{n+k} \to A^{n+k}, \quad t \in I$$

such that:

(i) $g_0 = g$;

(ii) there is a compact subset K of W such that $g_t|_{W \setminus K} = g|_{W \setminus K}$ for every $t \in I$;

(iii) $g_1^{-1}(M)$ is a closed PL submanifold of W, and the map $b :=$ $g_1|_{g_1^{-1}(M)} : g_1^{-1}(M) \to M$ is a homotopy equivalence.

We also say that the homotopy

$$G : W \times I \to A, \quad G(w,t) = g_t(w)$$

realizes the splitting of g.

(b) We say that g as in (a) is *transverally splittable* if the map g_1 from (a) is transversally regular to M.

An important special case is when $A^{n+k} = M^n \times B^k$ for some connected manifold B^k. In this case we can regard M as a submanifold $M \times \{b_0\}, b_0 \in$ B of A and say that $g : W \to A$ is splittable along M if it is splittable along $M \times \{b_0\}$. Clearly, this does not depend on the choice of $\{b_0\}$, i.e., g is splittable along $M \times \{b_0\}$ if and only if it is splittable along $M \times \{b\}$ with any other $b \in B$.

2.6.2 Lemma. *Let X, Y be two closed manifolds, $\dim X = \dim Y$, and let*

$$\varphi : (X \times \mathbb{R}, X \times \{0\}) \to (Y \times \mathbb{R}, Y \times \{0\})$$

be a map such that $\varphi^{-1}(Y \times \{0\}) = X \times \{0\}$. Then there is a map

$$\psi : (X \times \mathbb{R}, X \times \{0\}) \to (Y \times \mathbb{R}, Y \times \{0\})$$

that is homotopic to φ rel $X \times \{0\}$, ψ is transverse to $X \times \{0\}$, and

$$\psi^{-1}(Y \times \{0\}) = X \times \{0\}.$$

Proof. We write $\varphi(x,t) = (\varphi_1(x,t), \varphi_2(x,t)) \in Y \times \mathbb{R}$ where $\varphi_1(x,t)$ and $\varphi_2(x,t)$ are the projections of $\varphi(x,t)$ onto Y and $\mathbb{R}$, respectively. Now, put $\psi(x,t) = (\varphi_1(x,t), \lambda t + \varphi_2(x,t))$ with $\lambda \in \mathbb{R}$ large enough (λ should be larger than maximum absolute value of the "derivative with respect to t" of $\varphi_2(y,0)$ when y runs over Y). $\qquad \square$

2.6.3 Theorem. *Let $M^n, n \geq 5$ be a closed connected n-dimensional PL manifold such that $\pi_1(M)$ is a free Abelian group. Then every proper homotopy equivalence $h : W^{n+1} \to M^n \times \mathbb{R}$ is transversally splittable along M^n.*

Proof. Because of the Thom transversality theorem, there is a homotopy $h_t : W \to M \times \mathbb{R}$ satisfies condition (ii) of 2.6.1 and such that h_1 is transverse to M. We let $f = h_1$. Because of a crucial theorem of Novikov [N2,

Theorem 3], there is a sequence of surgeries of the inclusion $f^{-1}(M) \subset W$ in W such that the final result of these surgeries is a homotopy equivalence $V \subset W$. Using the Pontryagin–Thom construction, we realize this sequence of surgeries via a homotopy f_t such that f_t satisfies conditions (i)–(iii) of 2.6.1 with f_1 transverse to M and $f_1^{-1}(M) = V$. □

2.6.4 Theorem. *Let M^n be a manifold as in 2.6.3. Then every homotopy equivalence $h : W^{n+1} \to M^n \times S^1$ is transversally splittable along M^n.*

Proof. By [FH, Theorem 2.1], h is splittable along M. Indeed, since $\pi_1(M)$ is free Abelian, the obstruction $o(h)$ to splitting along M belongs to the group $\mathrm{Wh}(\mathbb{Z}^m) = 0$ (here Wh denote the Whitehead group). So, we get a homotopy equivalence $g : (W, W') \to (M \times S^1, M \times \{0\})$. Take a small $\varepsilon > 0$, denote the pair $(M \times (\varepsilon, \varepsilon), M \times \{0\})$ by $(Y \times \mathbb{R}, Y \times \{0\})$ and restrict g to a map

$$\varphi : (X \times \mathbb{R}, X \times \{0\}) \to (Y \times \mathbb{R}, Y)$$

with $X \times \{0\} = W'$ and $X \times \mathbb{R} \subset W$. Now, you get that h is transversally splittable by 2.6.2. □

2.6.5 Corollary. *Let M^n be a manifold as in 2.6.3. Let T^k denote the k-dimensional torus. Then every homotopy equivalence $W^{n+k} \to M^n \times T^k$ is transversally splittable along M^n.*

Proof. This follows from 2.6.4 by induction. □

2.6.6 Theorem. *Let M^n be a manifold as in 2.6.3. Then every homeomorphism $h : W^{n+k} \to M^n \times \mathbb{R}^k$ is transversally splittable along M^n.*

Proof. We use the Novikov's torus trick. Take a smooth inclusion

$$T^{k-1} \times \mathbb{R} \subset \mathbb{R}^k.$$

It yields the inclusion

$$M \times T^{k-1} \times \mathbb{R} \subset M \times \mathbb{R}^k.$$

We set $W_1 := h^{-1}(M \times T^{k-1} \times \mathbb{R})$. Note that W_1 is a smooth manifold since it is an open subset of W. Now, set

$$u = h|_{W_1} : W_1 \to M \times T^{k-1} \times \mathbb{R}. \tag{2.6.1}$$

Then, by 2.6.3, u is transversally splittable along $M \times T^{k-1}$, i.e., there is a homotopy u_t as in 2.6.1. We set $V := u_1^{-1}(M \times T^{k-1})$, and $g := f|_V$.

Because of 2.6.5, $g : V \to M \times T^{k-1}$ is splittable (in fact, even splits) along M. Hence, u_1 is transversally splittable along M, and therefore u in (2.6.1) is transversally splittable along M. This transverse splitting is realized by a homotopy $\bar{u}_t : W \times M \times \mathbb{R}^k$ with $\bar{u}_0 = u$. Now, we define the homotopy

$$k_t : W \to M \times \mathbb{R}^k$$

by setting

$$k_t|_{W_1} := \bar{u}_t|_{W_1} : W_1 \times I \to M \times T^{k-1} \times \mathbb{R} \to M \times \mathbb{R}^k$$

and $k_t|_{W \setminus W_1} := h|_{W \setminus W_1}$. Note that $\{k_t\}$ is a well-defined and continuous family since the family $\{\bar{u}_t\}$ satisfies 2.6.1(ii). It is clear that k_t realizes transverse splitting h along M. $\qquad \square$

2.6.7 Remarks. 1. The above Theorems 2.6.3, 2.6.4, and 2.6.6 of Novikov and Farrell–Hsiang were originally proved for smooth manifolds, but they are valid for PL manifolds as well, because there is an analogue of the Thom Transversality Theorem for PL manifolds, [Wil].

2. In the above-mentioned theorems we require the spaces to have free Abelian fundamental groups. For arbitrary fundamental groups, there are obstructions to the splittings that involves algebraic K-theory of the fundamental group π. In fact, in Theorem 2.6.4 there is an obstruction that is in an element of the Whitehead group $\mathrm{Wh}(\pi)$ of π. For Theorem 2.6.3, there are two obstructions: in $K^0(\pi)$ and in $\mathrm{Wh}(\pi)$.

2.6.8 Lemma. *Suppose that a map $g : W \to A$ is transversally splittable along a submanifold M of A. Let $\xi = \{E \to A\}$ be a PL bundle over A, let $g^*\xi = \{D \to W\}$, and let $\mathfrak{J}_g : g^*\xi \to \xi$ be the g-adjoint bundle morphism. Finally, let $l : D \to E$ be the map of the total spaces induced by $\mathfrak{J}_g$. Then l is transversally splittable along M. (Here the zero section of ξ allows us to regard A as a submanifold of E, and so M turns out to be a submanifold of E.)*

Proof. Let $G : W \times I \to A$ be a homotopy which realizes the splitting of g. Recall that $g^*\xi \times I$ is equivalent to $G^*\xi$. Now, the morphism

$$g^*\xi \times I \cong G^*\xi \xrightarrow{\ \mathfrak{J}_g\ } \xi$$

gives us the homotopy $D \times I \to E$ which realizes the splitting of l. $\qquad \square$

2.6.9 Lemma. *Let M be a manifold as in 2.6.3. Consider two PL $\mathbb{R}^d$-bundles $\xi = \{U \to M\}$ and $\eta = \{E \to M\}$ over M and a topological* *isomorphism $\varphi : \xi \to \eta$ over M of the form*

$$
\begin{array}{ccc}
U & \xrightarrow{\;g\;} & E \\
\downarrow & & \downarrow \\
M & =\!=\!= & M.
\end{array}
$$

Then there exists k such that the map

$$g \times 1 : U \times \mathbb{R}^k \to E \times \mathbb{R}^k$$

is splittable along M, where we regard M as a submanifold of E via the zero section of η.

Proof. Take a PL $\mathbb{R}^m$-bundle ζ such that $\eta \oplus \zeta = \theta^{d+m}$ and let W be the total space of $\xi \oplus \zeta$. Then the morphism

$$\varphi \oplus 1 : \xi \oplus \zeta \to \eta \oplus \zeta = \theta^{d+m}$$

yields a map of the total spaces

$$\Phi : W \to M \times \mathbb{R}^{d+m}. \tag{2.6.2}$$

Because of Theorem 2.6.6, the map Φ is splittable along M. Furthermore, the morphism

$$\varphi \oplus 1 \oplus 1 : \xi \oplus \zeta \oplus \eta \to \eta \oplus \zeta \oplus \eta$$

yields a map of the total spaces

$$g \times 1 : U \times \mathbb{R}^{2N+m} \to E \times \mathbb{R}^{2N+m}.$$

In view of Lemma 2.6.8, this map is splittable along M because Φ is. So, we can put $k = 2d + m$. $\qquad\square$

Now, let $a : TOP/PL \to G/PL$ be a map as in (1.3.7).

2.6.10 Theorem. *Let M be as in 2.6.3. Then the composition*

$$[M, TOP/PL] \xrightarrow{\;a_*\;} [M, G/PL] \xrightarrow{\;s\;} P_{\dim M}$$

is trivial, i.e., $sa_*(v) = 0$ for every $v \in [M, TOP/PL]$. In other words, $s(M, af) = 0$ for every $f : M \to TOP/PL$.

Proof. In view of (1.4.4), every element $v \in [M, TOP/PL]$ gives us a (class of a) topological morphism

$$\varphi : \nu_M^N \longrightarrow \gamma_{PL}^N$$

of PL $\mathbb{R}^N$-bundles with N large enough. Set $\xi = (\mathrm{bs}\,\varphi)^* \gamma_{PL}^N$. Then we get the adjoint morphism of PL $\mathbb{R}^N$-bundles $\mathfrak{I} : \xi \to \gamma_{PL}^N$. Take the correcting morphism $\psi : \nu_M \to \xi$ for φ. It is a commutative diagram

$$
\begin{array}{ccc}
A & \xrightarrow{\ g\ } & B \\
{\scriptstyle q}\downarrow & & \downarrow{\scriptstyle p} \\
M & =\!\!=\!\!= & M.
\end{array}
$$

However, now g is a homeomorphism because ψ is a topological isomorphism of PL bundles over M. So, by Lemma 2.6.9, the homeomorphism g splits over M (for N large enough). So, there is a homotopy $g_t : A \to B$ with $g_0 = g$ and g_1 is transverse to M, and $g_1 : V := g^{-1}(M) \to M$ is homotopy equivalence.

Recall that Thom spaces $T\nu, T\xi$ are the one-point compactifications of A and B, respectively. In particular, $T\psi$ is the one-point compactification Tg of g. Furthermore, there is a compact subset K of A such that $g_t(a) = g(a)$ for $a \notin K$. Hence, the homotopy g_t yields a homotopy $Tg_t : T\nu \to T\xi$.

Consider the reducible map

$$
h : S^{N+n} \xrightarrow{\ c\ } T\nu \xrightarrow{\ T\psi = T(g)\ } T\xi
$$

as in (2.3.2), where c is the collapsing map and $T\psi$ is a homeomorphism. Because of what we said above, we get a homotopy $h_t := cT(g_t)$ such that $h_1 : S^{N+n} \to T\xi$ is transverse to M and that $h_1 : h_1^{-1}(M) \to M$ is a homotopy equivalence. According to Definition 2.4.1, this means that $s(M, af) = 0$. $\qquad\square$

Now we show that the condition $\dim M \geqslant 5$ in 2.6.10 is not necessary.

2.6.11 Corollary. *Let M be a closed connected PL manifold such that $\pi_1(M)$ is a free Abelian group. Then $s(M, af) = 0$ for every map $f : M \to TOP/PL$.*

Proof. Let $\mathbb{CP}^2$ denote the complex projective plane, and let

$$
p_1 : M \times \mathbb{CP}^2 \to M
$$

be the projection on the first factor. Then $s(M \times \mathbb{CP}^2, gp_1) = s(M, g)$ for every $g : M \to G/PL$, see [Br2, Ch. III, §5]. In particular, for every map $f : M \to TOP/PL$ we have

$$
s(M, af) = s(M \times \mathbb{CP}^2, (af)p_1) = s(M \times \mathbb{CP}^2, a(fp_1)) = 0
$$

where the last equality follows from Theorem 2.6.10. $\qquad\square$

2.6.12 Remark. Probably, this is time to explain why we need the *transverse* splitting, not just splitting as in 2.6.1(a). The point is that, in the definiton of the invariant $s(M, f)$, we need to deform (2.3.2) to a *transverse* map $V \to M$. Otherwise, the normal bordism class is not well-defined.

2.7 Detecting Families

Recall the terminology: a singular smooth manifold in a space X is a map $M \to X$ of a smooth manifold M.

Given a CW-space X, consider a connected closed smooth singular orientable manifold $\gamma : M \to X$ in X. Then, for every map $f : X \to G/PL$, the invariant $s(M, f\gamma) \in P_{\dim M}$ is defined. Clearly, if f is inessential then $s(M, f\gamma) = 0$.

2.7.1 Definition. (a) Let $\{\gamma_j : M_j \to X\}_{j \in J}$ be a family of closed connected smooth singular manifolds in X; here J is an index set. We say that the family $\{\gamma_j : M_j \to X\}$ is a *detecting family* for X if, for every map $f : X \to G/PL$, the validity of all the equalities $s(M_j, f\gamma_j) = 0, j \in J$ implies that f is inessential.

(b) Given $m \in \mathbb{N}$, define the subset J_m of J by setting

$$J_m = \{j \in J | \dim M_j = m\}.$$

Note that G/PL is an H-space, and hence, for every detecting family $\{\gamma_j : M_j \to X\}$, the collection $\{s(M_j, f\gamma_j)\}$ determine a map $f : X \to G/PL$ uniquely up to homotopy.

The concept of detecting family is related to Sullivan's "characteristic variety", but it is not precisely the same. If a family $\mathcal{F}$ of singular manifolds in X contains a detecting family, then $\mathcal{F}$ on its own is a detecting family. On the contrary, the characteristic variety is, informally speaking, "minimal" detecting family.

2.7.2 Lemma. *Let X be a finite CW-space such that the group $H_*(X)$ is torsion free. Supppose that $\{\gamma_j : M_j \to X\}$ be a family of smooth oriented closed connected singular manifolds in X such that, for each m, the elements $(\gamma_j)_*[M_j^{2m}], j \in J_{2m}$ generate the group $H_{2m}(X)$. Then $\{\gamma_j\}$ is a detecting family for X.*

Proof. Consider a map $f : X \to G/PL$ such that $s_j(M_j, f\gamma_j) = 0$ for all $j \in J$. We must prove that f is inessential.

Because of 2.5.10, it suffices to prove that $f^*K_i = 0$ and $f^*K_{4i-2} = 0$ for all $i \geqslant 1$. Furthermore, $H^*(X) = \operatorname{Hom}(H_*(X), \mathbb{Z})$ because $H_*(X)$ is torsion free. So, it suffices to prove that

$$\langle f^*K_{4i}, z \rangle = 0 \text{ for all } z \in H_{4i}(X) \qquad (2.7.1)$$

and

$$\langle f^*K_{4i-2}, z \rangle = 0 \text{ for all } z \in H_{4i-2}(X; \mathbb{Z}/2). \qquad (2.7.2)$$

First, we prove (2.7.1). Since the classes $(\gamma_j)_*[M_j], \dim M_j = 4i$ with $j \in J_{4i}$ generate the group $H_{4i}(X)$, it suffices to prove that

$$\langle f^*K_{4i}, (\gamma_j)_*[M_j] \rangle = 0 \text{ whenever } \dim M_j = 4i.$$

But, because of 2.5.2, for every $4i$-dimensional M_i we have

$$0 = s(M_j, f\gamma_j) = \langle (f\gamma_i)^* K_{4i}, [M_j] \rangle = \langle f^*K_{4i}, (\gamma_j)_*[M_j] \rangle.$$

This completes the proof of the equality (2.7.1).

For the case $i = 4k - 2$, note that

$$H_*(X; \mathbb{Z}/2) = H_*(X) \otimes \mathbb{Z}/2$$

because $H_*(X)$ is torsion free. Hence, the group $H_{4i-2}(X; \mathbb{Z}/2)$ is generated by the elements

$$(\gamma_j)_*[M_j] \otimes 1 \in H_{4i-2}(X) \otimes \mathbb{Z}/2 \cong H_{4i-2}(X; \mathbb{Z}/2), \quad \dim M_j = 4j - 2.$$

Now the proof can be completed similarly to the case $i = 4k$. $\qquad \square$

2.7.3 Theorem. *Let X be a connected finite CW-space such that the group $H_*(X)$ is torsion free. Then X admits a detecting family $\{\gamma_j : M_j \to X\}$ such that each M_j is orientable.*

Proof. Since $H_*(X)$ is torsion free, every homology class in $H_*(X)$ can be realized by a closed connected smooth oriented singular manifold, see e.g., [Co, 15.2] or [Rud, IV.6.6 and IV.7.36]. Now apply Lemma 2.7.2. $\qquad \square$

2.7.4 Example. Let X be the space $T^n \times S^k$. Clearly, $H_{2m}(X)$ is generated by fundamental classes of submanifolds T^{2m} and $T^{2m-k} \times S^k$ of $T^n \times S^k$. Hence, X has a detected family $\{\gamma_j : M_j \to X\}$ such that each M_j is either T^r or $T^r \times S^k$.

2.8   Normal Invariant of a Homeomorphism $V \to T^n \times S^k$

2.8.1 Theorem. *If the element* $x \in \mathcal{S}_{PL}(T^n \times S^k), k + n \geqslant 5$ *can be represented by a homeomorphism* $h : V \to T^k \times S^n$, *then* $j_G(x) = 0$.

Proof. Put $M = T^n \times S^k$. The maps j_{TOP} and j_G from section 1.4 can be incorporated in the commutative diagram

$$
\begin{array}{ccc}
\mathcal{T}_{PL}(M) & \xrightarrow{\ j_{TOP}\ } & [M, TOP/PL] \\
\downarrow & & \downarrow{\scriptstyle a_*} \\
\mathcal{S}_{PL}(M) & \xrightarrow{\ j_G\ } & [M, G/PL]
\end{array}
\qquad (2.8.1)
$$

where the left arrow is the forgetful map (1.4.5) and a_* is induced by a as in (1.3.7).

Suppose that x can be represented by a homeomorphism $h : V \to M$. Consider a map $f : M \to TOP/PL$ such that $j_{TOP}(h)$ is a homotopy class of f. Then, clearly, the class $j_G(x) \in [M, G/PL]$ is represented by the map

$$
M \xrightarrow{\ f\ } TOP/PL \xrightarrow{\ a\ } G/PL.
$$

As we explained in Example 2.7.4, M possesses a detecting family $\{\gamma_j : M_j \to M\}$ where each M_j is either T^r or $T^r \times S^n$. Hence, by 2.6.10 and 2.6.11, $s(M_j, af\gamma_j) = 0$ for all j. So, af is inessential since $\{\gamma_j\}$ is a detecting family. Thus, $j_G(x) = 0$. $\square$

Chapter 3

Applications and Consequences of the Main Theorem

3.1 The Space G/TOP

Because of the Main Theorem and results of Freedman [F], Scharle-mann [Sch], and Quinn [FQ, Q3], the Transversality Theorem holds for *topological* manifolds and bundles. (See also Rudyak [Rud, IV.7.18] for a bit more detailed references.)

Since we have the topological transversality, we can define the maps

$$s' : [M, G/TOP] \to P_{\dim M}$$

where M turns out to be a topological manifold. These map s' are obvious analog of maps s defined in Definition 2.4.1: you just have to replace PL by TOP in Definition 2.4.1. We leave it to the reader.

To see the important difference between G/PL and G/TOP, com-pare Theorem 2.4.3(iii) with the following Proposition 3.1.1. See also Re-mark 3.1.5 below.

3.1.1 Proposition. *The map*

$$s' : \pi_4(G/TOP) = [S^4, G/TOP] \to \mathbb{Z}$$

is a surjection.

Proof. Note that the Freedman *topological* manifold V^4 from Theorem 1.8.2 is almost parallelizable and has signature 8. Because of this, for *topological* manifolds the argument from Theorem 2.4.3(i) holds for $i = 1$ as well. In detail, we apply Proposition 2.3.10 to topological manifolds and get a normal morphism φ with the base $V^4 \to S^4$. Because of (a topological analog of) Theorem 2.3.9, we have $\tilde{\Gamma}[\varphi] = [f]$ for a certain map $f : S^4 \to G/TOP$. Thus,

$$s(S^4, f) = \frac{\sigma(V^4) - \sigma(S^4)}{8} = 1.$$

Now the proof can be completed as 2.4.3(i).   □

3.1.2 Remark. Siebenmann [Sieb4] used a *homology* 4-manifold of signature 8 in order to prove 3.1.1 and forthcoming 3.1.3(ii). The paper of Freedman [F] appeared later.

Recall the fibration

$$TOP/PL \xrightarrow{\ a\ } G/PL \xrightarrow{\ b\ } G/TOP$$

from (1.3.7).

3.1.3 Theorem. (i) *For $i \neq 4$ the map $b : G/PL \to G/TOP$ induces an isomorphism*

$$b_* : \pi_i(G/PL) \longrightarrow \pi_i(G/TOP).$$

(ii) *The homomorphism*

$$b_* : \mathbb{Z} = \pi_4(G/PL) \longrightarrow \pi_4(G/TOP) = \mathbb{Z}$$

is the multiplication by 2.

Proof. (i) Recall that $TOP/PL \simeq K(\mathbb{Z}/2, 3)$ and $\pi_4(G/PL) = \mathbb{Z}$. So, exactness of the homotopy sequence of the fibration

$$TOP/PL \xrightarrow{\ a\ } G/PL \xrightarrow{\ b\ } G/TOP$$

yields an isomorphism $b_* : \pi_i(G/PL) \cong \pi_i(G/TOP)$ for $i \neq 4$.

(ii) We have the commutative diagram

$$0 = \pi_4(TOP/PL)$$

$$a_* \downarrow$$

$$\mathbb{Z} = \pi_4(G/PL) \xrightarrow{\ s\ } \mathbb{Z}$$

$$b_* \downarrow$$

$$\pi_4(G/TOP) \xrightarrow{\ s'\ } \mathbb{Z}$$

$$\downarrow$$

$$\mathbb{Z}/2 = \pi_3(TOP/PL)$$

$$\downarrow$$

$$0 = \pi_3(G/PL)$$

where the (middle) vertical line is a short exact sequence. Therefore $\pi_4(G/TOP) = \mathbb{Z}$ or $\pi_4(FTOP) = \mathbb{Z} \oplus \mathbb{Z}/2$. By Theorem 2.4.3(iii), Im s is the subgroup $2\mathbb{Z}$ of $\mathbb{Z}$, while s' is a surjection by 3.1.1. Thus, $\pi_4(G/TOP) = \mathbb{Z}$ and b_* is the multiplication by 2.   □

Now, following 2.5.2, we can introduce the classes

$$K'_{4i} \in H^{4i}(G/TOP; \mathbb{Z}[2]) \text{ and } K'_{4i-2} \in H^{4i-2}(G/TOP; \mathbb{Z}/2)$$

such that

$$s'(M^{4i}, f) = \langle f^* K'_{4i}, [M] \rangle \text{ and } s'(N^{4i-2}, f) = \langle f^* K'_{4i-2}, [N]_2 \rangle.$$

However, here M and N are allowed to be *topological* (i.e., not necessarily PL) manifolds.

Similarly to (2.5.2), together these classes yield the map

$$K' : G/TOP \longrightarrow \prod_{i>0} (K(\mathbb{Z}[2], 4i) \times K(\mathbb{Z}/2, 4i - 2))$$

such that for each $i > 0$ the map

$$G/TOP \xrightarrow{\ K'\ } \Pi \xrightarrow{\ \text{projection}\ } K(\mathbb{Z}/2, 4i - 2) \text{ (resp. } K(\mathbb{Z}[2], 4i))$$

coincides with K'_{4i-2} (resp. K'_{4i}).

3.1.4 Theorem. *The map*

$$K'[2] : G/TOP[2] \longrightarrow \prod_{i>0} (K(\mathbb{Z}[2], 4i) \times K(\mathbb{Z}/2, 4i - 2))$$

is a homotopy equivalence.

Proof. Together 2.4.3 and 3.1.1 imply that the homomorphisms

$$s' : \pi_{2i}(G/TOP) \longrightarrow P_{2i}$$

are surjective. Now, in view of 3.1.3, all these homomorphisms are isomorphisms, and the result follows.   □

3.1.5 Remark. Now you see that the spaces G/PL and G/TOP have isomorphic homotopy groups. Moreover, the only (small but crucial) difference between the spaces G/PL and G/TOP is that all Postnikov invariants of $G/TOP[2]$ are trivial, while $G/PL[2]$ has exactly one non-trivial Postnikov invariant $\delta Sq^2 \iota_2 \in H^5(K(\mathbb{Z}/2, 2); \mathbb{Z}[2])$.

Now we discuss the difference between $\pi_i(BPL)$ and $\pi_i(BTOP)$. Consider the map

$$\alpha = \alpha_{TOP}^{PL} : BPL \to BTOP$$

and the fibration

$$TOP/PL \longrightarrow BPL \xrightarrow{\alpha} BTOP$$

as in (1.3.6). Since $\pi_3(BPL) = \pi_3(BO) = 0$ and $\pi_i(TOP/PL) = 0$ for $i \neq 3$, we conclude that

$$\alpha_* : \pi_i(BPL) \longrightarrow \pi_i(BTOP)$$

is an isomorphism for $i \neq 4$. Furthermore, we have the exact sequence

$$0 \longrightarrow \pi_4(BPL) \xrightarrow{\alpha_*} \pi_4(BTOP) \longrightarrow \pi_3(TOP/PL) \longrightarrow 0$$

where $\pi_4(BPL) = \mathbb{Z}$ and $\pi_3(TOP/PL) = \mathbb{Z}/2$. Hence, we see that either $\pi_4(BTOP = \mathbb{Z}$ or $\pi_4(BTOP) = \mathbb{Z} \oplus \mathbb{Z}/2$.

Now, consider the diagram of fibrations

$$
\begin{array}{ccccc}
G/PL & \longrightarrow & BPL & \longrightarrow & BG \\
\downarrow & & \downarrow & & \| \\
G/TOP & \longrightarrow & BTOP & \longrightarrow & BG
\end{array}
\qquad (3.1.1)
$$

It is known that J-homomorphism

$$J : \mathbb{Z} = \pi_4(BPL) \longrightarrow \pi_4(BSF) = \mathbb{Z}/24$$

is surjective [Ad, MK] (since $\pi_i(PL/O) = 0$ for $i < 7$, there is no difference between $\pi_i(BPL)$ and $\pi_i(BO)$ up to dimension 6). Furthermore, $\pi_5(BG)$ is finite (in fact, $\pi_5(BG) = 0$, but we do not need to use this equality) and $\pi_3(G/PL) = \pi_3(G/TOP) = 0$. Now, we apply π_4 to the diagram (3.1.1) and get the commutative diagram with exact rows

$$
\begin{array}{ccccccccc}
0 & \longrightarrow & \mathbb{Z} & \xrightarrow{24} & \mathbb{Z} & \longrightarrow & \mathbb{Z}/24 & \longrightarrow & 0 \\
& & {\scriptstyle 2}\downarrow & & \downarrow & & \downarrow{\scriptstyle =} & & \\
0 & \longrightarrow & \mathbb{Z} & \longrightarrow & \pi_4(BTOP) & \longrightarrow & \mathbb{Z}/24 & \longrightarrow & 0.
\end{array}
$$

The assertion $\pi_4(BTOP) = \mathbb{Z}$ contradicts the commutativity of the diagram. Thus, $\pi_4(BTOP) = \mathbb{Z} \oplus \mathbb{Z}/2$, see Milgram [Mil].

3.2   The Map $a : TOP/PL \to G/PL$

Recall that in (1.3.7) we described the fibration

$$TOP/PL \xrightarrow{\;a\;} G/PL \xrightarrow{\;b\;} G/TOP.$$

3.2.1 Proposition. *The map* $a : TOP/PL \to G/PL$ *is essential.*

Proof. The fibration

$$TOP/PL \xrightarrow{\;a\;} G/PL \longrightarrow G/TOP$$

yields the fibration

$$\Omega(G/TOP) \xrightarrow{\;u\;} TOP/PL \xrightarrow{\;a\;} G/PL,$$

see e.g., [MT]. If a is inessential then there exists a map

$$v : TOP/PL \to \Omega(G/TOP)$$

with $uv \simeq 1$. But this is impossible because $\pi_3(TOP/PL) = \mathbb{Z}/2$ while $\pi_3(\Omega(G/TOP)) = \pi_4(G/TOP) = \mathbb{Z}$. $\Box$

Let $\ell : G/PL \to G/PL[2]$ denote the localization map. Let $\psi : G/PL \to Y$ be the Postnikov 4-approximation of G/PL as in (2.5.3). Let X be a CW space of finite type. Take any map $f : X \to TOP/PL$.

3.2.2 Proposition. *The following three assertions are equivalent:*

    (i) *the map*

$$X \xrightarrow{\;f\;} TOP/PL \xrightarrow{\;a\;} G/PL$$

is essential;

    (ii) *the map*

$$X \xrightarrow{\;f\;} TOP/PL \xrightarrow{\;a\;} G/PL \xrightarrow{\;\ell\;} G/PL[2]$$

is essential;

    (iii) *the map*

$$X \xrightarrow{\;f\;} TOP/PL \xrightarrow{\;a\;} G/PL \xrightarrow{\;\ell\;} G/PL[2] \xrightarrow{\;\psi[2]\;} Y[2]$$

is essential.

Proof. It suffices to prove that (i) $\Rightarrow$ (ii) $\Rightarrow$ (iii). To prove the first implication, recall that a map $af : X \to G/PL$ is inessential if both localized maps

$$X \xrightarrow{\;af\;} G/PL \to G/PL[2], \qquad X \xrightarrow{\;af\;} G/PL \to G/PL[1/2]$$

are inessential. Now, (i) $\Rightarrow$ (ii) holds since $TOP/PL[1/2]$ is contractible.

To prove the second implication, note that a map $\ell af : X \to G/PL[2]$ is inessential if both maps (we use notation as in Section 2.5)

$$X \xrightarrow{\ell af} G/PL[2] \xrightarrow{K[2]} \Pi, \qquad X \xrightarrow{\ell af} G/PL[2] \xrightarrow{\psi[2]} Y[2]$$

are inessential. Furthermore, the map

$$X \xrightarrow{\ell af} G/PL[2] \xrightarrow{K[2]} \Pi \qquad\qquad (3.2.1)$$

is inessential. Indeed, the commutative diagram

$$
\begin{array}{ccccc}
G/PL[2] & \xrightarrow{K[2]} & \Pi & = & \Pi \\
\downarrow{\scriptstyle b[2]} & & & & \| \\
G/TOP[2] & \xrightarrow{K'[2]} & \prod_{i>0}(K(\mathbb{Z}/2, 4i-2) \times K(\mathbb{Z}[2], 4i)) & \xrightarrow{\text{proj}} & \Pi
\end{array}
$$

produces the commutative diagram

$$
\begin{array}{ccccc}
X & \xrightarrow{\ell af} & G/PL[2] & \xrightarrow{K[2]} & \Pi \\
\| & & \downarrow{\scriptstyle b[2]} & & \| \\
X & \xrightarrow{\alpha\ell af} & G/TOP[2] & \xrightarrow{\text{proj}\circ K'[2]} & \Pi.
\end{array}
$$

But the map $\alpha\ell af : X \to G/TOP[2]$ passes through the inessential map $TOP/PL \to G/PL \to G/TOP$ and therefore is inessential.

Now, since the map (3.2.1) is inessential, we conclude that the map $\ell af : X \to G/PL[2]$ is inessential iff the map

$$X \xrightarrow{\ell af} G/PL[2] \xrightarrow{\psi[2]} Y[2]$$

is. Thus, (ii) and (iii) are equivalent. $\square$

Consider the fibration

$$K(\mathbb{Z}[2], 4) \xrightarrow{\;i\;} Y[2] \longrightarrow K(\mathbb{Z}/2, 2)$$

that is the $\mathbb{Z}[2]$-localization of the fibration (2.5.4).

3.2.3 Lemma. *For every space X, the homomorphism*

$$H^4(X; \mathbb{Z}[2]) = [X, K(\mathbb{Z}[2], 4)] \xrightarrow{\;i_*\;} [X, Y[2]]$$

is injective. Moreover, i_* is an isomorphism if $H^2(X; \mathbb{Z}/2) = 0$.

Proof. The fibration (2.5.4) yields the exact sequence (see e.g., [MT])

$$H^1(X;\mathbb{Z}/2) \xrightarrow{\delta Sq^2} H^4(X;\mathbb{Z}[2]) \xrightarrow{i_*} [X,Y[2]] \to H^2(X;\mathbb{Z}/2)$$

where $\delta Sq^2 = 0$ (because $Sq^2(x) = 0$ whenever $\deg x = 1$). $\qquad\square$

Let $g : TOP/PL \to Y$ be the composition

$$TOP/PL \xrightarrow{\;a\;} G/PL \xrightarrow{\;\ell\;} G/PL[2] \xrightarrow{\;\psi[2]\;} Y[2].$$

Note that g is essential because of Propositions 3.2.1 and 3.2.2.

3.2.4 Corollary. *The map*

$$TOP/PL \simeq K(\mathbb{Z}/2,3) \xrightarrow{\;\delta\;} K(\mathbb{Z}[2],4) \xrightarrow{\;i\;} Y[2]$$

is homotopic to g, *i.e.,* $g \simeq i\delta$.

Proof. Because of Lemma 3.2.3 applied to $X = K(\mathbb{Z}/2,3)$, we have an isomorphism

$$H^4(K(\mathbb{Z}/2,3);\mathbb{Z}[2]) \xrightarrow{\;i_*\;} [K(\mathbb{Z}/2,3),Y[2]].$$

So, $i_*(\delta) \neq 0$ since $\delta \in H^4(K(\mathbb{Z}/2,3);\mathbb{Z}[2]) \neq 0$. In other words, $i \circ \delta$ is essential. Furthermore, the set $[K(\mathbb{Z}/2,3),Y[2]]$ has exactly two elements since $H^4(K(\mathbb{Z}/2,3);\mathbb{Z}[2])$ does. Since both maps g and $i \circ \delta$ are essential, we conclude that $g \simeq i\delta$. $\qquad\square$

3.2.5 Corollary. *Given a map* $f : X \to TOP/PL$, *the map*

$$X \xrightarrow{\;f\;} TOP/PL \xrightarrow{\;a\;} G/PL$$

is essential if and only if the map

$$X \xrightarrow{\;f\;} TOP/PL \simeq K(\mathbb{Z}/2,3) \xrightarrow{\;\delta\;} K(\mathbb{Z}[2],4)$$

is essential.

Proof. We have the chain of equivalences

$$af \text{ is essential } \xleftrightarrow{\;3.2.2\;} gf \text{ is essential } \xleftrightarrow{\;3.2.4\;} i\delta f \text{ is essential}$$

$$\xleftrightarrow{\;3.2.3\;} \delta f \text{ is essential.}$$

$\qquad\square$

3.3 Normal Invariant of a Homeomorphism

3.3.1 Lemma. *Let X be a finite CW-space such that $H_n(X)$ is 2-torsion free. Then the homomorphism*

$$\delta : H^n(X;\mathbb{Z}/2) \to H^{n+1}(X;\mathbb{Z}[2])$$

is zero.

Proof. Because of exactness of the sequence (2.5.5)

$$H^n(X;\mathbb{Z}/2) \xrightarrow{\ \delta\ } H^{n+1}(X;\mathbb{Z}[2]) \xrightarrow{\ 2\ } H^{n+1}(X;\mathbb{Z}[2]),$$

it suffices to prove that $H^{n+1}(X;\mathbb{Z}[2])$ is 2-torsion free. Since $H_n(X)$ is 2-torsion free, we conclude that $\mathrm{Ext}(H_n(X),\mathbb{Z}[2]) = 0$. (Indeed, $\mathrm{Ext}(\mathbb{Z}/m,A) = A/mA$ for all A.) Thus, because of the Universal Coefficient Theorem,

$$H^{n+1}(X;\mathbb{Z}[2]) \cong \mathrm{Hom}(H_{n+1}(X),\mathbb{Z}[2]) \oplus \mathrm{Ext}(H_n(X),\mathbb{Z}[2])$$
$$= \mathrm{Hom}(H_{n+1}(X),\mathbb{Z}[2]),$$

and the result follows.   □

3.3.2 Theorem (Sullivan [Sul2]). *Let M be a closed PL manifold such that $H_3(M)$ is 2-torsion free. Then the normal invariant of any homeomorphism $h : V \to M$ is trivial.*

Proof. Let $j_{TOP}(h) \in [M, TOP/PL]$ be represented by a map $f : M \to TOP/PL$, see 1.5.1. Then the normal invariant $j_G(h) \in [M, G/PL]$ is represented by a map

$$M \xrightarrow{\ f\ } TOP/PL \xrightarrow{\ a\ } G/PL.$$

We have to prove that $j_G(h) = 0$, i.e., that af is inessential. Because of Corollary 3.2.5, it suffices to prove that

$$M \xrightarrow{\ f\ } TOP/PL \simeq K(\mathbb{Z}/2,3) \xrightarrow{\ \delta\ } K(\mathbb{Z}[2],4)$$

is inessential. To do this, it suffices, in turn, to prove that

$$\delta : H^3(M;\mathbb{Z}/2) \to H^4(M;\mathbb{Z}[2])$$

is the zero homomorphism. But this follows from Lemma 3.3.1.   □

Now we have the following version of the *Hauptvermutung*, cf. [Cas, Corollary on p. 68] and [Sul2, Theorem H on p. 93].

3.3.3 Corollary. *Let* $M, \dim M \geqslant 5$ *be a closed simply-connected PL manifold such that* $H_3(M)$ *is 2-torsion free. Then every homeomorphism* $h : V \to M$ *is homotopic to a PL homeomorphism. In particular, V and M are PL homeomorphic.*

Proof. This follows from 2.4.2 and 3.3.2. In greater detail, take a homeomorphism $h : V \to M$ and let $[h] \in \mathcal{S}_{PL}(M)$ denote the equivalence class of h. We have $j_G[h] = 0$ because of Theorem 3.3.2. But $j_G : \mathcal{S}_{PL} \to [M, G/PL]$ is injective by Theorem 2.4.2. So, $[h]$ is the trivial element of $\mathcal{S}_{PL}(M)$. In other words, there is a PL homeomorphism $\varphi : V \to M$ such that the diagram

$$
\begin{array}{ccc}
V & \xrightarrow{\ \varphi\ } & M \\
\ \downarrow h & & \ \downarrow 1_M \\
M & =\!=\!= & M
\end{array}
$$

commutes up to homotopy. Thus, $h : V \to M$ is homotopic to the PL isomorphism φ. ☐

3.3.4 Remark. Rourke [Rou] suggested another proof of 3.3.2, using the technique of simplicial sets.

3.3.5 Remark. Lashof and Rothenberg [LR2] proved a weak version of 3.3.3: Let $h : Q \to M$ be a topological homeomorphism between two closed PL manifolds of dimension at least five. If M is 4-connected, then h is homotopic to a PL homeomorphism. See also [Arm].

3.4 Kirby–Siebenmann and Casson-Sullivan Invariants

Recall some facts on obstruction theory [DK, FFG, H, MT, Spa2]. Let

$$ F \to E \to B $$

be a principal F-fibration such that F is an Eilenberg-MacLane space $K(\pi, n)$, and assume that the $\pi_1(B)$-action on $\pi = \pi_n(F)$ is trivial. Let

$$ \iota = \iota_n \in H^n(K(\pi, n); \pi) $$

be the fundamental class of F, and let

$$ \kappa = \tau\iota \in H^{n+1}(B; \pi) $$

be the characteristic class of the fibration $F \to E \to B$ where

$$\tau : H^n(F; \pi) \to H^{n+1}(B; \pi)$$

is the transgression. This is well known that the fibration $F \to E \to B$ admits a section if and only if $\kappa = 0$ and, if a section exists then the vertically homotopy classes of sections of the fibration correspond bijectively with elements of $H^n(B; \pi)$. Hence, given a map $f : X \to B$, the map f can be lifted to E iff $f^*(\kappa) = 0$, and the vertical homotopy classes of liftings of f to E are in a bijective correspondence with elements $H^n(X; \pi)$ provided that such a lifting exists.

Since TOP/PL is the Eilenberg-MacLane space $K(\mathbb{Z}/2, 3)$, we can apply previous arguments to the principal TOP/PL-fibration

$$TOP/PL \longrightarrow BPL \xrightarrow{a_{TOP}^{PL}} BTOP$$

described in (1.3.6).

3.4.1 Definition. Let $\iota \in H^3(K(\mathbb{Z}/2, 3); \mathbb{Z}/2)$ be the fundamental class. Define the *universal Kirby–Siebenmann invariant* as the characteristic class

$$\varkappa = \tau \iota \in H^4(BTOP; \pi_3(K(\mathbb{Z}/2, 3))) = H^4(BTOP; \mathbb{Z}/2)$$

of the fibration (1.3.6).

3.4.2 Definition. Let M be a topological manifold, and let $f : M \to BTOP$ classify the stable tangent bundle of M. Since f is unique up to homotopy, the class $f^*(\varkappa) \in H^4(M; \mathbb{Z}/2)$ is a well-defined invariant of M. We put

$$\varkappa(M) := f^*(\varkappa) \in H^4(M; \mathbb{Z}/2)$$

and call it the *Kirby–Siebenmann invariant of M*, or *Kirby–Siebenmann class of M*.

3.4.3 Theorem. *Let M be a topological manifold. If M admits a PL structure then $\varkappa(M) = 0$. If $\dim M \geqslant 5$ and $\varkappa(M) = 0$ then M admits a PL structure. In particular, if $\dim M \geqslant 5$ and $H^4(M; \mathbb{Z}/2) = 0$ then M admits a PL structure.*

Proof. If M admits a PL structure then the classifying map $f : M \to BTOP$ can be lifted to BPL, and hence $f^*(\varkappa) = 0$, i.e., $\varkappa(M) = 0$. Conversely, if $\varkappa(M) = 0$ then f can be lifted to BPL. Thus, in case $\dim M \geqslant 5$ the manifold M admits a PL structure by Corollary 1.7.4. $\qquad\square$

3.4.4 Theorem. *If a topological manifold* $M, \dim M \geqslant 5$ *admits a PL structure then the set of concordance classes of PL structures on* M *corresponds bijectively with* $H^3(M; \mathbb{Z}/2)$, *i.e.,*

$$\mathcal{T}_{PL}(M) \cong H^3(M; \mathbb{Z}/2).$$

In particular, if $H^3(M; \mathbb{Z}/2) = 0$ *then the* Hauptvermutung *holds for* M.

Proof. Corollary 1.7.2 implies a bijection

$$\mathcal{T}_{PL}(M) \cong [M, TOP/PL].$$

Thus, because of the Main Theorem $TOP/PL \simeq K(\mathbb{Z}/2, 3$, we get

$$\mathcal{T}_{PL}(M) \cong [M, TOP/PL] \cong [M, K(\mathbb{Z}/2, 3] \cong H^3(M; \mathbb{Z}/2).$$

$\square$

3.4.5 Definition. Let M be a topological manifold, V be a PL manifold and $h : V \to M$ be a topological homeomorphism (i.e., a PL structure on M). In view of bijection from Theorem 3.4.4, the PL structure h gives us a cohomology class

$$\varkappa(h) = \varkappa_M(h) \in H^3(M; \mathbb{Z}/2).$$

This class is called the *Casson–Sullivan invariant* of h, and it measures the difference between $h : V \to M$ and 1_M.

So, $\varkappa(h) = 0$ if and only if $h : V \to M$ is concordant to the identity map of M. It is also worthy to mention that, for every $a \in H^3(M; \mathbb{Z}/2)$, there exists a homeomorphism $h : V \to M$ with $a = \varkappa(h)$.

3.4.6 Remark. The following interpretation of Casson–Sullivan invariant $\varkappa(h)$ looks interesting. Consider the mapping cylinder of h,

$$W = (V \times [0,1] \cup M/) \sim, \text{where } (v,1) \sim h(v), \ v \in V.$$

Now, W is a topological manifold, whose boundary $\partial W = V \sqcup M$ is a PL manifold. Hence, we have a map $W \to BTOP$ whose boundary is lifted to BPL. So, we can try the extend the PL structure on ∂W over the whole W. The obstruction to this extension is the relative Kirby–Siebenmann invariant $\varkappa(W, \partial W)$. Now, we have

$$\varkappa(h) = \varkappa(W, \partial W) \in H^4(W, \partial W; \mathbb{Z}/2) \cong H^3(M; \mathbb{Z}/2).$$

In other words, Casson–Sullivan invariant turns out to be the relative Kirby–Siebenmann invariant.

3.4.7 Remark. We know that Hauptvermutung holds for $T^n \times S^k$ with $k + n \geqslant 5$ and $k \geqslant 4$, [HS]. In other words, if two PL manifolds M_1, M_2 are homeomorphic to $T^n \times S^k$ then there are PL homeomorphic. On the other hand, the group $H^3(T^n \times S^k; \mathbb{Z}/2)$ is quite large for n large enough, i.e., $T^n \times S^k$ has many different PL structures. Is it a contradiction? No, it is not. The explanation comes because, given a homeomorphism $h : T^n \times S^k \to T^n \times S^k$, there are many PL concordance classes $T^n \times S^k \to T^n \times S^k$ that are homotopic to h.

3.5 Several Examples

3.5.1 Example. *There are two closed PL manifolds that are homeomorphic but not PL homeomorphic.*

Let $\mathbb{RP}^n$ denote the real projective space of dimension n and assume that $n > 4$.

There are a PL manifold M and a (topological) homeomorphism $k : \mathbb{RP}^n \to M$ such that

$$0 \neq j_{TOP}(k) \in [M, TOP/PL] = H^3(M; \mathbb{Z}/2) = \mathbb{Z}/2.$$

Since M is homeomorphic to $\mathbb{RP}^n$, the Bockstein homomorphism

$$\beta : \mathbb{Z}/2 = H^3(M; \mathbb{Z}/2) \to H^4(M; \mathbb{Z}/2) = \mathbb{Z}/2$$

is an isomorphism. Hence $\beta(j_{TOP}(k)) \neq 0$, and so $\delta(j_{TOP}(k)) \neq 0$ where

$$\delta : H^3(M; \mathbb{Z}/2) \to H^4(M)$$

is the coboundary homomorphism as in (2.5.5). So, by Theorem 3.2.5, $a_* j_{TOP}(k) \neq 0$. In view of commutativity of the diagram (2.8.1) we have $j_G(k) = a_* j_{TOP}(k)$, i.e., $j_G(k) \neq 0$.

We claim that M is not PL homeomorphic to $\mathbb{RP}^n$. Suppose the contrary, and take a PL homeomorphism $\varphi : \mathbb{RP}^n \to M$. First, consider the case when n is odd.

Consider a PL involution $T : \mathbb{RP}^n \to \mathbb{RP}^n$ of degree -1. Recall that every self-equivalence $\mathbb{RP}^n \to \mathbb{RP}^n$ with n odd is homotopic to either $1_{\mathbb{RP}^n}$ or T, [Gor]. So, the map $k : \mathbb{RP}^n \to M$ is homotopic to either φ or φT. Hence, the map k is homotopic to a PL homeomorphism, i.e., k yields a trivial element of $\mathcal{S}_{PL}(M)$. Thus, $j_G(k) = 0$. That is a contradiction.

In case of n even, note that every self-equivalence $\mathbb{RP}^n \to \mathbb{RP}^n$ is homotopic to the identity map. Now the proof can be proved similarly to (and simpler as) the case of n odd.

3.5.2 Remark. Originally, in March 1969 Siebenmann disproved the *Hauptvermutung* for manifolds by constructing two PL manifolds that are homeomorphic to the torus $T^n, n \geqslant 5$ but not PL homeomorphic to each other. See [Sieb4] for a publication. Note also that the failure of the *Hauptvermutung* allowed Kirby [K1] to prove the annulus conjecture in dimension ≥ 5.

3.5.3 Example. *For every $n > 3$ there is a homeomorphism*

$$h = h_n : S^3 \times S^n \to S^3 \times S^n$$

which is not concordant to the identity map.

Take an arbitrary homeomorphism $f : M \to S^3 \times S^n, n > 3$. Then $j_G(f)$ is trivial by Theorem 3.3.2. Thus, by Theorem 2.4.2, f is homotopic to a PL homeomorphism. In particular, M is PL homeomorphic to $S^3 \times S^n$.

We refine the situation and take a homeomorphism

$$h : M \to S^3 \times S^n$$

such that

$$j_{TOP}(h) \neq 0 \in \mathcal{T}_{PL}(S^3 \times S^n) = H^3(S^3 \times S^n; \mathbb{Z}/2) = \mathbb{Z}/2.$$

Such h exists because j_{TOP} is a bijection. So, h is not concordant to the identity map. But, as we have already seen, M is PL homeomorphic to $S^3 \times S^n$.

Note that the maps h and the identity map have the same domain while they are not concordant. So, this example serves also the Remark 1.4.5.

3.5.4 Examples. *There are topological manifolds that do not admit any PL structure.*

See manifold $V \times T^n$ that are described in Corollary 1.8.4. See also Remark 1.8.6.

In 1970 Siebenmann [Sieb2] published a paper with the intriguing title: *Are nontriangulable manifolds triangulable?* The paper cerebrated about the following problem: Are there manifolds that can be triangulated as simplicial complexes but do not admit any PL structure? Later, people made good progress related to this issue, [AM, GaSt1, GaSt2, Mat, Man, Rand, Sav]. Here we give only a brief survey notice on the issue because non-combinatorial triangulations are far from the main line of our concern.

Recall that a homology k-sphere is defined to be a k-dimensional closed PL manifold Σ such that $H_*(\Sigma) \cong H_*(S^k)$.

3.5.5 Examples. *There are topological manifolds that can be triangulated as simplicial complexes but do not admit any PL structure.*

3.5.6 Theorem. *Every orientable topological 5-dimensional closed manifold can be triangulated as a simplicial complex.*

Proof. Let us say that a homology 3-sphere Σ is *good* if the double suspension $S^2\Sigma$ over Σ is homeomorphic to S^5 and Σ bounds a compact parallelizable manifold of signature 8. Siebenmann [Sieb2, Assertion on p. 81] proved that every orientable topological 5-dimensional closed manifold can be triangulated as a simplicial complex provided that there exists a good homology 3-sphere. Cannon [Ca] proved that, for any homology 3-sphere Σ, the double suspension $S^2\Sigma$ is homeomorphic to S^5. Now, note that the homology 3-sphere ∂W from Comment 1.8.3 is good, and the theorem follows.   □

Let V be the Freedman's manifold descrided in 1.8.2. Take $M = V \times S^1$. Then M does not admit any PL structure by Corollary 1.8.4. On the other hand, M can be triangulated as a simplicial complex by Theorem 3.5.6. Because of this, for each $k \geqslant 1$ the manifold $V \times T^k$ also have these properties. This is remarkable that V cannot be triangulated as a simplicial complex, see below.

3.5.7 Examples. *There are topological manifolds that cannot be triangulated as simplicial complexes.*

First, note that if a 4-dimensional topological manifold N can be triangulated as a simplicial complex then A admits a PL structure. In particular, V cannot be triangulated as a simplicial complex (Casson), see [AM, Sav].

Now we pass to higher dimensions.

Define two oriented homology 3-spheres Σ_1, Σ_2 to be *equivalent* if there exists an oriented PL bordism W, $\partial W = \Sigma_1 \sqcup \Sigma_2$ such that $H_1(W) = 0 = H_2(W)$. Let Θ_3^H denote the Abelian group obtained from the set of equivalence classes using the operation of connected sum. We define a homomorphism $\mu : \Theta_3^H \to \mathbb{Z}/2$ as follows.

By a theorem of Rokhlin, every orientable 3-manifold bounds a 4-dimensional parallelizable manifold, see [K2]. Take a homology 3-sphere Σ and consider a 4-dimensional parallelizable manifold P with $\partial P = \Sigma$. By the Rokhlin Theorem 1.8.1, the signature $\sigma(P)$ mod 16 is a well-defined invariant of Σ, and 8 divides $\sigma(P)$. (Indeed, consider two membranes P_1, P_2

with $\partial P_1 = \Sigma = \partial P_2$ and note that $w_i(P_1 \cup_\Sigma P_2) = 0$ for $i = 1, 2$.) Take $a \in \Theta_3^H$, let Σ_a be a homology 3-sphere that represents a, and let P_a be a 4-dimensional parallelizable manifold with $\partial P_a = \Sigma_a$. Now, put

$$\mu(a) = \frac{\sigma(P_a)}{8} \mod 16$$

and get a well-defined homomorphism $\mu : \Theta_H^3 \to \mathbb{Z}/2$. Note that μ is surjective: Indeed, $\mu(\Sigma) = 1$ where Σ denote the Poincare homology 3-sphere (the dodecahedral space).

Consider the short exact sequence

$$0 \longrightarrow \ker \mu \overset{\subset}{\longrightarrow} \Theta_3^H \overset{\mu}{\longrightarrow} \mathbb{Z}/2 \longrightarrow 0$$

and let $\delta : H^4(-; \mathbb{Z}/2) \to H^5(-; \ker \mu)$ be the coboundary homomorphism associated with this sequence.

3.5.8 Theorem (Galewski–Stern [GaSt2], Matumoto [Mat]). *A topological manifold M of dimensional $\geqslant 5$ can be triangulated as a simplicial complex if and only if $\delta\varkappa(M) = 0$. Here $\varkappa(M)$ denotes the Kirby–Siebenmann invariant of M.*

Manolescu [Man] proved that the above mentioned short exact sequence does not split. This allowed him to prove that, for any $n \geqslant 5$, there is a manifold M^n with $\delta\varkappa(M^n) \neq 0$. Thus, for all $n \geqslant 5$ there exists an n-dimensional manifold that cannot be triangulated as a simplicial complex.

Concerning explicit constructions of such manifolds. Galewski and Stern [GaSt1] constructed a certain manifold N^5 with the following property: if N can be triangulated as a simplicial complex then *every* closed manifold of dimension $\geqslant 5$ can. So, N cannot be triangulated. In particular, $\delta\varkappa(N) \neq 0$. Finally, $N \times T^k$ cannot be triangulated as a simplicial complex because $\delta\varkappa(N \times T^k) \neq 0$.

Summary

Here all manifolds are assumed to be connected and having the homotopy type of a finite CW complex.

1. *Every manifold M^n with $n \leqslant 3$ admits a PL structure that is unique up to PL homeomorphism (trivial assertion for $n = 1$, Rado [Rad] (for triangulability) and Papakyriakopoulos [P] (for uniqueness) for $n = 2$, Moise [Mo] for $n = 3$).*

2. *There are uncountable set of mutually different PL manifolds that are homeomorphic to $\mathbb{R}^4$ (Taubes [Ta], cf. also [GS, K2]). There are countably*

infinite set of mutually different closed 4-dimensional PL manifolds that are homeomorphic to the blow-up of $\mathbb{CP}^2$ *at the nine points of intersection of two general cubics (Okonek–Van de Ven [OV]).*

3. *For every* $n \geqslant 5$ *there exist a pair of closed* n-*dimensional PL manifolds that are homeomorphic but not PL homeomorphic to each other. So, the Hauptvermutung is wrong in general. However, any topological manifold* $M^n, n \geqslant 5$ *(not necessarily closed) possesses only finite number of (concordance classes of) PL structures (Kirby–Siebenmann [KS2]).*

4. *For every* $n \geqslant 4$ *there exist closed topological* n-*dimensional manifolds that do not admit any PL structure (Freedman [F] for* $n = 4$, *Kirby–Siebenmann [KS2] for* $n > 4$).

The item 4 can be bifurcated as follows:

4a. *For every* $n \geqslant 5$ *there exists an* n-*manifold that does not possess any PL structure but can be triangulated as a simplicial complex (Siebenmann [Sieb2] + Cannon [Ca]). Such examples do not exist for* $n \leqslant 4$.

4b. *For every* $n \geqslant 4$ *there exists an* n-*manifold that cannot be triangulated as a simplicial complex (Casson [AM, Sav] for* $n = 4$, *Manolescu [Man] for* $n \geqslant 5$).

3.6   Topological and Homotopy Invariance of Characteristic Classes

Given a real vector bundle ξ over a space X, the kth *Pontryagin class* of ξ is a cohomology class $p_k(\xi) \in H^{4k}(X)$, [MS]. In particular, for every smooth manifold M we have the Pontryagin classes $p_k(M) := p_k(\tau M)$ where τM is the tangent bundle of M. Given a commutative ring Λ with unit, we can consider $p_k(\xi) \in H^{4k}(X; \Lambda)$, the image of the Pontryagin class $p_k(\xi) \in H^{4k}(X)$ under the coefficient homomorphism $\mathbb{Z} \to \Lambda$. In particular, we have *rational Pontryagin classes* $p_k(\xi) \in H^{4k}(X; \mathbb{Q})$ and *modulo* p *Pontryagin classes* $p_k(\xi) \in H^{4k}(X; \mathbb{Z}/p)$.

In this section we discuss homotopy and topological invariance of some characteristic classes. In particular, we prove that the Novikov's Theorem [N2] on topological invariance of rational Pontryagin classes is a direct corollary of the Main Theorem. (It is worthy to note, however, that the proof of the Main Theorem uses ideas from [N2].) Concerning other proofs of the Novikov's theorem see [G, ST, RW].

3.6.1 Definition. Given a class $x \in H^*(BO; \Lambda)$, we say that x is *topologically invariant* if, for any two maps $f_1, f_2 : B \to BO$ such that

$$\alpha^O_{TOP} f_1 \simeq \alpha^O_{TOP} f_2 : B \to BTOP,$$

we have

$$f_1^*(x) = f_2^*(x) \text{ in } H^*(B; \Lambda).$$

Now we give some conditions for topological invariance. Similarly to the fibration (1.3.6), consider the fibration

$$TOP/O \xrightarrow{\beta} BO \xrightarrow{\alpha} BTOP.$$

3.6.2 Proposition. (i) *If*

$$x \in \operatorname{Im}\{\alpha^* : H^*(BTOP; \Lambda) \to H^*(BO; \Lambda)\}$$

then x is topologically invariant. In particular, if Λ is such that α^* is epimorphic then every class $x \in H^*(BO; \Lambda)$ is topologically invariant.

(ii) *If $x \in H^*(BO; \Lambda)$ is topologically invariant then $\beta^*(x) = 0$ for $\beta^* : H^*(BO) \to H^*(TOP/O)$.*

Proof. (i) is obvious. To prove (ii), note that $\alpha\beta$ is inessential. Hence $\alpha\beta \simeq \alpha\varepsilon$ where $\varepsilon : TOP/O \to BO$ is a constant map. Since x is topologically invariant, we conclude that $\beta^*(x) = \varepsilon^*(x) = 0$. $\qquad\square$

Proposition 3.6.2(i) tells us a sufficient condition for topological invariance, while 3.6.2(ii) tells us a necessary condition. We will see below that 3.6.2(i) is not necessary and 3.6.2(ii) is not sufficient for topological invariance. Now we give a condition that is necessary and sufficient at the same time. Consider the map

$$\mu : BO \times TOP/O \xrightarrow{1 \times \beta} BO \times BO \xrightarrow{m} BO$$

where m is the multiplication in the H-space BO.

3.6.3 Theorem. *The class $x \in H^*(BO; \Lambda)$ is topologically invariant if and only if $\mu^*(x) = x \otimes 1 \in H^*(BO; \Lambda) \otimes H^*(TOP/O; \Lambda)$.*

Proof. The map $\alpha\beta$ is topologically trivial, and hence $\alpha\mu$ is homotopic to the map

$$\alpha\nu : BO \times TOP/O \to BO \to BTOP$$

where $\nu : BO \times TOP/O \to BO$ is the projection on the first factor. Since x is topologically invariant, we conclude that $\mu^* x = \nu^*(x) = x \otimes 1$.

Conversely, suppose that $\mu^*(x) = x \otimes 1$. Recall that, for all X, the infinite space structure in BO turns $[X, BO]$ into an Abelian group. Let $f_1, f_2 : B \to BO$ be two maps such that $\alpha f_1 \simeq \alpha f_2$. Recall that $[X, BO]$ is an Abelian group with respect to the infinite space structure in BO. Then $f_2 - f_1 : B \to BO$ lifts to a map $B \to TOP/O$. In other words, $f_2 = f_1 + g$ for some $g : B \to TOP/O$. Hence we have a homotopy commutative diagram

$$
\begin{array}{ccc}
B & \xrightarrow{\;f_2\;} & BO \\
{\scriptstyle\Delta}\big\downarrow & & \big\uparrow{\scriptstyle\mu} \\
B \times B & \xrightarrow{\;f_1 \times g\;} & BO \times TOP/O
\end{array}
$$

Now

$$
\begin{aligned}
f_2^*(x) &= \Delta^*(f_1 \times g)^* \mu^*(x) = \Delta^*(f_1 \times g)^*(x \otimes 1) = \Delta^*(f_1^*(x) \otimes 1) \\
&= f_1^*(x).
\end{aligned}
$$

$\square$

3.6.4 Remark. The items 3.6.1–3.6.3 are taken from the paper of Sharma [S].

The following lemma plays a crucial role for topological invariance of rational Pontryagin classes.

3.6.5 Lemma. *The forgetful map* $\alpha_{TOP}^O : BO[0] \to BTOP[0]$ *is a homotopy equivalence. Thus, the forgetful map* $\alpha_{TOP}^O : BO \to BTOP$ *induces an isomorphism*

$$
(\alpha_{TOP}^O)^* : H^*(BTOP; \mathbb{Q}) \to H^*(BO; \mathbb{Q}).
$$

Proof. First, note that the homotopy groups $\pi_i(PL/O)$ are finite, see [Rud, IV.4.27(iv)] for the references. Hence, the space $PL/O[0]$ is contractible. Thus, $\alpha_{PL}^O : BO[0] \to BPL[0]$ is a homotopy equivalence.

Second, the homotopy groups $\pi_i(TOP/PL)$ are finite by the Main Theorem. Hence, the space $TOP/PL[0]$ is contractible. Thus, $\alpha_{TOP}^{PL} : BPL[0] \to BTOP[0]$ is a homotopy equivalence.

Now, since $\alpha_{TOP}^O = \alpha_{TOP}^{PL} \alpha_{PL}^O$, we conclude that $\alpha_{TOP}^O[0]$ is a homotopy equivalence. $\square$

Recall that $H^*(BO; \mathbb{Q}) = \mathbb{Q}[p_1, \ldots, p_i, \ldots]$ where $p_k, \dim p_k = 4k$ is the universal Pontryagin class, [MS]. It follows from Lemma 3.6.5 that

$H^*(BTOP; \mathbb{Q}) = \mathbb{Q}[p_1', \ldots, p_k', \ldots]$ where p_k' are the cohomology classes determined by the condition

$$\alpha^*(p_k') = p_k \in H^*(BO; \mathbb{Q}).$$

Now, given an arbitrary topological $\mathbb{R}^n$-bundle λ over B, we define its rational Pontryagin classes $p_k'(\lambda) \in H^{4i}(B; \mathbb{Q})$ by setting

$$p_k'(\lambda) = t^* p_k'$$

where $t : B \to BTOP$ classifies λ.

3.6.6 Theorem. *Every class in $H^*(BO; \mathbb{Q})$ is topologically invariant. In other words, if $\xi_i = \{\pi_i : E_i \to B\}, i = 1, 2$ are two topologically isomorphic vector bundles over a space B then $p_k(\xi_1; \mathbb{Q}) = p_k(\xi_2; \mathbb{Q})$.*

This is the famous Novikov theorem on topological invariance of rational Pontryagin classes.

Proof. This follows from Lemma 3.6.5 and Proposition 3.6.2(i) immediately.
$\square$

For completeness, we state the original Novikov version of topological invariance, see [N2].

3.6.7 Theorem. *Let $f : M_1 \to M_2$ be a homeomorphism of closed smooth manifolds, and let $f^* : H^*(M_2; \mathbb{Q}) \to H^*(M_1; \mathbb{Q})$ be the induced isomorphism. Then $f^* p_k(M_2; \mathbb{Q}) = p_k(M_1; \mathbb{Q})$ for all k.*

Proof. Let $t_s : M_s \to BO \to BTOP, s = 1, 2$ classify the stable tangent bundle of M_s. Then $t_1 \simeq t_2 f$. Now

$$f^* p_k(M_2; \mathbb{Q}) = f^* t_2^* p_k' = (t_2 f)^* p_k' = t_1^* p_k' = p_k(M_1; \mathbb{Q}),$$

and we are done.
$\square$

3.6.8 Remark. We can pass the previous issues to PL category. To define PL invariance, we should replace topological isomorphism by PL isomorphism of PL bundles and require B to be a polyhedron in Definition 3.6.1. Rokhlin and Švarc [RoS] and Thom [T] proved PL invariance of rational Pontryagin classes in 1957-58th. Of course, this result follows from the Novikov Theorem 3.6.7 on topological invariance of rational Pontryagin classes, but the Novikov Theorem appeared almost 10 years later.

So, rational Pontryagin classes are topological invariants. What about *integral* Pontryagin classes? It turns out to be that they are not even PL invariant. Milnor [Mi4, §9] constructed two smooth manifolds M_1, M_2 that are PL homeomorphic while $p_2(M_1) = 0, p_2(M_2) \neq 0$ (and $7p_2(M_2) = 0$).

Nevertheless, there are certain topological invariance results for integral Pontryagin classes.

3.6.9 Notation. Because of Lemma 3.6.5, the index of the image subgroup

$$\text{Im}\{(\alpha_{TOP}^{O})^* : H^m(BTOP) \to H^m(BO)\}$$

in $H^m(BO)$ is finite for each m. Let ε_k denote this index for $m = 4k$. Clearly, the class $\varepsilon_k p_k \in H^{4k}(BO)$ (the multiple of the integral Pontryagin class) is topologically invariant.

Define $e_k \in \mathbb{N}$ to be the smallest number such that $e_k p_k$ is topologically invariant.

3.6.10 Comment. To evaluate e_k, Sharma [S] proved the following. Let d_k be the smallest positive integer such that

$$d_k p_k \in \text{Ker}\{\beta* : H^*(BO) \to H^*(TOP/O)\}.$$

Then $e_k = \text{LCM}(d_1, \ldots, d_k)$. In particular, $e_k | e_{k+1}$. To compute d_k, let

$$\gamma_k = (2^{2k-1} - 1) \text{ numerator } (B_{2k}/4k).$$

Here B_m's are the Bernoulli numbers, [Wash]. Now, if p is an odd prime which divides γ_k but does not divide γ_i with $i < k$, then $\nu_p(d_k) = \nu_p(\gamma_k)$. Here, as usual, $m = p^{\nu_p(m)} a$ with $(a, p) = 1$.

Sharma [S, Theorem 1.6] used these results in order to evaluate e_k for $k \leqslant 8$. In particular, $e_1 = 1$, $e_2 = 7$, $e_3 = 7 \cdot 31$, $e_4 = 7 \cdot 31 \cdot 127$. It is remarkable to note the *strict* inequality $e_4 < \varepsilon_4$, [S, Prop. 1.7 ff]. So, there are topologically invariant classes that do not come from $BTOP$, i.e., the sufficient condition 3.6.2(i) for topological invariance is not necessary. To see that the necessary condition 3.6.2(ii) is not sufficient, note that $31p_3$ is not topological invariant because e_3 does not divide 31, while $31p_3 \in \text{Ker}\,\beta$, see [S, Section 4, proof of Theorem 1.3].

Another kind of topological invariance appears when we consider p_k mod m, the modulo m Pontryagin classes. Here we will not give detailed proofs but give a sketch/survey only. As a first example, note that p_k mod $2 = w_{2k}^2$, and hence p_k mod 2 is topologically (and even homotopy) invariant in view of homotopy invariance of Stiefel–Whitney classes, [MS]. So, the question about topological invariance of modulo p Pontryagin classes is not vacuous. In fact, we have the following result, see [Si, SS]:

3.6.11 Theorem. *Given an odd prime p, let $n(p)$ be the smallest value of k such that p divides e_k. Then p_k mod p is a topological invariant for $k < n(p)$ and is not a topological invariant for $k \geqslant n(p)$. In particular, if p does not divide e_k for all $k \geqslant 1$, then p_k mod p is a topological invariant.*

Because of Theorem 3.6.11 and Comment 3.6.10, one can prove that the classes p_k mod p are topologically invariant for all k and $p = 3, 5, 11, 13, 17$, while p_k mod 7 is not a topological invariant. (For $p = 3$ it is an old theorem of Wu, see Theorem 3.6.14.)

Now some words about homotopy invariance.

3.6.12 Definition. Given a class $x \in H^*(BO; \Lambda)$, we say that x is *homotopy invariant* if, for any two maps $f_1, f_2 : B \to BO$ such that

$$\alpha_F^O f_1 \simeq \alpha_F^O f_2 : B \to BG,$$

we have

$$f_1^*(x) = f_2^*(x) \text{ in } H^*(B; \Lambda).$$

The obvious analogs of Proposition 3.6.2 and Theorem 3.6.3 remains valid if we speak about homotopy invariance instead of topological invariance and replace TOP by G.

3.6.13 Proposition. *The non-zero rational Pontryagin classes are* not *homotopy invariant.*

Proof. Note that $\pi_i(BG)$ is isomorphic to the stable homotopy group π_{i-1}^S and therefore is finite because of a well-known theorem of Serre, [Se]. Hence, $\pi_i(BG) \otimes \mathbb{Q} = 0$, and so $BG[0]$ is contractible. Now consider the fibration $G/O \to BO \to BG$ and conclude that $\beta[0] : G/O[0] \to BO[0]$ is a homotopy equivalence, and hence $\beta^* : H^*(BO; \mathbb{Q}) \to H^*(G/O; \mathbb{Q})$ is an isomorphism. Thus, because of the homotopy analog of 3.6.2(ii), we see that $x \in \widetilde{H}^*(BO; \mathbb{Q})$ is homotopy invariant iff $x = 0$.   □

On the other hand, we have $p_i \mod 2 = w_{2i}^2$, [MS], i.e., $p_i \mod 2$ is a homotopy invariant. So, it seems reasonable to ask about homotopy invariance of $p_i \mod p$ for odd prime p.

Recall that the homotopy invariance of Stiefel-Whitney follows from the Thom-Wu formula $w_i(\xi) = \varphi^{-1} Sq^i u$ where u is the Thom class of ξ and

$$\varphi : H^*(B; \mathbb{Z}/2) \to \widetilde{H}^{*+n}(T\xi; \mathbb{Z}/2)$$

is the Thom isomorphism, [MS]. (Here $T\xi$ is the Thom space of the $\mathbb{R}^n$-bundle ξ over B.)

We apply this idea modulo p. So, let p be an odd prime and

$$\mathcal{P}^k : H^*(-; \mathbb{Z}/p) \to H^{*+2k(p-1)}(-; \mathbb{Z}/p)$$

be the Steenrod power, [St1]. Given an oriented $\mathbb{R}^n$-bundle (or an $(S^n, *)$-fibration) ξ over B, let $T\xi$ be the Thom space of ξ, let $u \in H^n(T\xi; \mathbb{Z}/p)$ be the Thom class, and let

$$\varphi : H^*(B; \mathbb{Z}/p) \to \widetilde{H}^{*+n}(T\xi; \mathbb{Z}/p)$$

be the Thom isomorphism. Then

$$q_k(\xi) := \varphi^{-1}\mathcal{P}^k(u) \in H^{2k(p-1)}(X)$$

is a characteristic class, and it is homotopy invariant by construction. For $X = BO$ we get a universal characteristic class q_k, and it is a polynomial of universal Pontryagin classes mod p. Wu [Wu] proved that $q_k = p_k$ if $p = 3$. So, we get the following theorem.

3.6.14 Theorem (Wu). *The Pontryagin classes p_k mod $3, k \geqslant 1$ are homotopy invariant.*

Madsen [M] proved that the classes p_k mod $8, k \geqslant 1$ are homotopy invariant. So, we have the following result:

3.6.15 Corollary. *The classes p_k mod $24, k \geqslant 1$ are homotopy invariant.*

Appendix

Quinn's Proof of Product Structure Theorem

In this appendix we apply the theory of ends of spaces in order to prove the Product Structure Theorem 1.7.1. The exposition here is taken from the papers of Quinn [Q1, Q2, Q4]. We need some preliminaries.

Given an open manifold V, it is natural to ask if one can find a boundary for V. In other words, to find a compact manifold $\widehat{V}$ such that V is homeomorphic (PL homeomorphic, diffeomorphic) to the interior of $\widehat{V}$. In [BLL] the authors found the necessary and sufficient conditions for the existence of boundary for simply-connected V, Siebenmann [Sieb1] generalized the situation for non-simply connected manifolds.

Quinn [Q1, Q2] posed a more general question. Given an open manifold M and a map $h : M \to X$, can we extend h to a proper map $\widehat{h} : \widehat{M} \to X$ by adding boundary to M? Quinn found the conditions that ensure the existence of $\widehat{h}$. As it is expected, for $X = \mathrm{pt}$ the Quinn's theorem turns into the above-mentioned results from [BLL, Sieb1]. We need to cite some definitions from [Q1, Q2].

A.1 Definition. Let V be an open manifold and let $h : V \to X$ be a map. A *completion* of h is a proper map $\widehat{h} : \widehat{V} \to X$ where $\widehat{V}$ is a manifold with boundary such that $V \subset \widehat{V}$, $\widehat{V} \setminus V \subset \partial \widehat{V}$, and $\widehat{h}|_V = h$.

A.2 Definition. Let $h : V \to X$ be as in Definition A.1.

   (a) Define an open subset U of V to be *good* if the map

$$h|_{(V \setminus U)} : V \setminus U \to X$$

is proper.

(b) The map h is *tame* if, for every good open set U and continuous function $\varepsilon : X \to (0, \infty)$ there are a good open set $W \subset U$ and a homotopy $H : V \times I \to V$ such that

$$H_0 = 1_V, \; H_1(V) \subset V \setminus W, \; H_t(V \setminus U) \subset V \setminus W \text{ for all } t \in [0,1],$$

and that for each $v \in V$ the diameter of the arc $hH : \{v\} \times I \to X$ is less than ε.

(c) The map h is 1-LC (locally contractible in dimension 1) if for every $x \in X$, neighborhood $W \subset X$ of x, and every good open set U, there are neighborhood $W' \subset W$ and a good open set $U' \subset U$ such that points in $U' \cap h^{-1}(W')$ can be joined by arcs in $U \cap h^{-1}(W)$ and, moreover, loops in $U' \cap h^{-1}(W')$ are contractible in $U \cap h^{-1}(W)$.

A.3 Theorem. *Let X be a locally compact locally simply-connected metric space. Let $V, \dim V \geqslant 6$ be an open manifold. If $h : V \to X$ is a tame and 1-LC such that $h(U) = X$ for every good open set $U \subset V$, then h admits a completion.*

Proof. See [Q2, Theorem 1.4]. □

A.4 Definition. A continuous map $f : V \to M$ of topological manifolds is called a *CE map* if f is proper and, for each $y \in M$ and each neighborhood U of $f^{-1}y$ in M there exists a (smaller) neighborhood U' of $f^{-1}y$ such that the inclusion $U' \subset U$ is homotopic to a constant map.

This definition is formally different from that in [Sieb3, §1], but they coincides for manifolds.

A.5 Theorem. *Let $f : V \to M$ be a CE map of metric topological m-manifolds without boundary, $m \geqslant 5$. Let $\varepsilon : V \to (0, \infty)$ be a continuous map. Then there exists a continuous family $f_t : V \to M, t \in [0,1]$ such that $f_0 = f$ and, for $0 < t \leqslant 1$, the map f_t is a homeomorphism with $\operatorname{dist}(f(x), f_t(x)) < \varepsilon(x)$.*

Proof. See [Sieb3, Approximation Theorem A and Complement to Theorem A, pp. 271–272]. □

Now we pass to the Product Structure Theorem 1.7.1. Recall that the theorem claims that the map

$$e : \mathcal{T}_{PL}(M) \to \mathcal{T}_{PL}(M \times \mathbb{R}^k), \quad h \mapsto h \times 1,$$

is a bijection for $\dim M \geqslant 5$. First, it is sufficient to prove the bijectivity of $e : \mathcal{T}_{PL}(M) \to \mathcal{T}_{PL}(M \times \mathbb{R})$, the general case follows by the obvious induction (iteration). Furthermore, we prove only the surjectivity of e, the injectivity can be proved in the same manner, see [Q4, p. 306].

It is convenient for us to regard $M \times \mathbb{R}$ as the homeomorphic space $M \times (0, 1)$.

So, given a PL manifold M, $\dim M \geqslant 5$ and a homeomorphism $f : V \to M \times (0, 1)$, we should find a PL manifold N and a homeomorphism $h : N \to M$ such that $h \times 1 : N \times (0, 1) \to M \times (0, 1)$ is concordant to f.

Consider the map

$$\widetilde{h} : V \xrightarrow{\ h\ } M \times (0, 1) \xrightarrow{\ \subset\ } M \times (0, 1].$$

Because of Theorem A.3, there exists a completion $\widehat{h} : \widehat{V} \to M \times (0, 1]$ of $\widehat{h}$. (It is clear that the conditions of Theorem A.3 are fulfilled, we leave it to the reader to check it.) In particular, $\widehat{h}$ is a proper map such that $\widehat{h}_{|V} = h$ and $\widehat{h}^{-1}(M \times (0, 1)) = V$. Furthermore, $\widehat{h}^{-1}(M \times \{1\}) \subset \partial \widehat{V}$. We put $\partial_\infty \widehat{V} = \widehat{h}^{-1}(M \times \{1\})$ and define $\partial_\infty \widehat{h} : \partial_\infty \widehat{V} \to M \times \{1\}$ by setting $(\partial_\infty \widehat{h})(x) = \widehat{h}(x)$. (So, $\partial_\infty \widehat{h}$ is the restriction of $\widehat{h}$.)

Note that $\partial_\infty \widehat{h} : \partial_\infty \widehat{V} \to M \times \{1\}$ is a CE map since $h : V \to M \times (0, 1)$ is a homeomorphism of manifolds. Hence, by Theorem A.5, there is a family

$$f_t : \partial_\infty \widehat{V} \to M \times \{1\}, \quad t \in [0, 1]$$

such that $f_0 = \partial_\infty \widehat{h}$ and f_t is a homeomorphism for $t > 0$. Now, we regard $f_1 : V \to M$ as a PL structure on M, and the map

$$F : V \times (0, 1) \to M \times (0, 1), \quad F(v, t) = f_t(v), \, v \in V, \, t \in (0, 1)$$

is concordant to $f_1 \times 1$.

Bibliography

[Ad]   J. F. Adams, *On the groups J(X), II.* Topology **3** (1965) pp. 137–171.

[AM]   S. Akbulut and J. McCarthy, An Exposition. *Cassons invariant for oriented homology 3-spheres*, Mathematical Notes, vol. **36** (Princeton University Press, Princeton, NJ, 1990).

[AH]   P. Alexandroff and H. Hopf, *Topologie. I.* Berichtigter Reprint. (Reprint of the 1935 first edition.) Die Grundlehren der mathematischen Wissenschaften, Band **45** (Springer-Verlag, Berlin-New York, 1974).

[Arm]   M. Armstrong, *The Hauptvermutung according to Lashof and Rothenberg.* In: The Hauptvermutung book (the item [Ran]), pp. 107–127.

[Astey]   L. Astey, *Commutative 2-local ring spectra.* Proc. R. Soc. Edinb., Sect. A **127**, No.1 (1997) pp. 1–10.

[Atiyah] M. Atiyah, *Thom complexes*, Proc. London Math. Soc. (3) **11** (1961) pp. 291–310.

[BV]   J. Boardman and R. Vogt, *Homotopy invariant algebraic structures*, Lecture Notes in Mathematics **347** (Springer, Berlin, 1972).

[Br1]   W. Browder, *Homotopy type of differentiable manifolds.* Proceedings of the Aarhus Symposium (1962), pp. 42–46.

[Br2]   W. Browder, *Surgery on simply-connected manifolds*, Ergebnisse der Mathematik 5 (Springer, Berlin, 1972).

[BLL]   W. Browder, J. Levine, and G. Livesay, *Finding a boundary for an open manifold*, Amer. J. Math. **87** (1965) no. 4, pp. 101–1028.

[Cai]   S. Cairns, *Triangulation of the manifold of class one*, Bull. Amer. Math. Soc. **41** (1935) no. 8, pp. 549–552.

[Ca]   J. Cannon, *Shrinking cell-like decompositions of manifolds. Codimension three*, Ann. of Math. (2) **110** (1979) no. 1, pp. 83–112.

[Cas]   A. Casson, *Generalisations and Applications of Block Bundles*, In: The Hauptvermutung book (the item [Ran]), pp. 33–68.

[Ch]   A. V. Černavski, *Local contractibility of the homeomorphism group of a manifold*, Math. U.S.S.R Sbornik **8** (1968) no. 3, 287–333.

[C]   M. Cohen, *A course in simple-homotopy theory*, Graduate Texts in Mathematics **10** (Springer, Berlin, 1973).

[Co]   P. Conner, *Differentiable periodic maps* (second edition), Lecture Notes in Mathematics **738** (Springer, Berlin, 1979).

[DK]   J. Davis and P. Kirk, *Lecture notes in algebraic topology*, Graduate Studies in Mathematics, **35** (Amer. Math. Soc., Providence, RI, 2001).

[Dold]   A. Dold, *Partitions of unity in the theory of fibrations*, Ann. of Math. (2) **78** (1963) pp. 223–255.

[DP]   A. Dold and D. Puppe, *Duality, trace, and transfer*, Proceedings of the International Conference on Geometric Topology, Warsaw, 1978 (PWN, Warsaw, 1980), pp. 81–102.

[EK]   R. Edwards and R. Kirby, *Deformations of classes of embeddings*, Ann. of Math. **93** (1971) pp. 63–88.

[FH]   F.T. Farrell and W.C. Hsiang, *Manifolds with $\pi_1 = G \times_\alpha T$*, Amer. J. Math. **95** (1973) pp. 813–848.

[FFG]   A. Fomenko, D. Fuchs, and V. Gutenmacher, *Homotopic topology* (Hungarian Academy of Sciences, Budapest, 1986).

[FU]   D. Freed and K. Uhlenbeck, *Instantons and four-manifolds*, Mathematical Sciences Research Institute Publications, 1 (Springer-Verlag, New York-Berlin, 1984).

[F]   M. Freedman, *The topology of four-dimensional manifolds*, J. Differential Geom. **17** (1982) no. 3, pp. 357–453.

[FQ]   M. Freedman and F. Quinn, *Topology of 4-manifolds*, Princeton Mathematical Series, **39** (Princeton University Press, Princeton, NJ, 1990).

[FR]   D. B. Fuks and V. A. Rokhlin, *Beginners course in topology. Geometric chapters*, (Springer Series in Soviet Mathematics, Springer-Verlag, Berlin, 1984).

[GaSt1] D. Galewski and R. Stern, *A universal 5-manifold with respect to simplicial triangulations*, Geometric topology (Proc. Georgia Topology Conf., Athens, GA, 1977), pp. 345–350 (Academic Press, New York-London, 1979).

[GaSt2] D. Galewski and R. Stern, *Classification of simplicial triangulations of topological manifolds*, Ann. of Math. (2) **111** (1980) no. 1, pp. 1–34.

[GS]   R. Gompf and A. Stipsicz, *4-manifolds and Kirby calculus*, Graduate Studies in Mathematics **20** (Amer. Math. Soc., Providence, RI, 1999).

[Gor]   I. Gordon, *Classification of the mappings of an n-dimensional complex into an n-dimensional real projective space* (Russian) Izvestiya Akad. Nauk SSSR. Ser. Mat. **16** (1952) pp. 113–146.

[Gray]   B. Gray, *Homotopy theory. An introduction to algebraic topology*, (Academic Press [Harcourt Brace Jovanovich, Publishers], New York-London, 1975).

[G]   M. Gromov, *Positive curvature, macroscopic dimension, spectral gaps and higher signatures*, Functional analysis on the eve of the 21st century, Vol. II (New Brunswick, NJ, 1993), pp. 1–213, Progr. Math., 132, (Birkhäuser Boston, Boston, MA, 1996).

[HW]   A. Haefliger and C. T. C. Wall, *Piecewise linear bundles in the stable range*, Topology **4** (1965) pp. 209–214.

[Hat]   A. Hatcher, *Algebraic topology* (Cambridge University Press, Cambridge, 2002).

[HMR]   P. Hilton, G. Mislin and J. Roitberg, *Localization of nilpotent groups and spaces*, North-Holland Mathematics Studies **15**, Notas de Matematica **55** (North-Holland, Amsterdam-Oxford, 1975).

[H]   M. Hirsch, *On non-linear cell bundles*, Ann. of Math. (2) **84** (1966) pp. 373–385.

[HM]   M. Hirsch and B. Mazur *Smoothings of piecewise linear manifolds*, Annals of Mathematics Studies **80** (Princeton University Press, Princeton, 1974).

[Hirz]   F. Hirzebruch, *Topological methods in algebraic geometry*, Die Grundlehren der Mathematischen Wissenschaften, Band **131** (Springer-Verlag New York, Inc., New York, 1966).

[HS]   W. Hsiang and J. Shaneson, *Fake Tori*, Topology of Manifolds, Athens Georgia Conf., 1969 (eds. J.C. Cantrell and C.H. Edwards, Markham, Chicago, 1970), pp. 18–51.

[Hud]   J. Hudson, *Piecewise linear topology*, University of Chicago Lecture Notes prepared with the assistance of J. L. Shaneson and J. Lees (W. A. Benjamin, Inc., New York-Amsterdam 1969).

[Hus]   D. Husemoller, *Fibre bundles*. Third edition. Graduate Texts in Mathematics, **20** (Springer-Verlag, New York, 1994).

[KM]   M. Kervaire and J. Milnor, *Groups of homotopy spheres I*, Ann. of Math. (2) **77** (1963) pp. 504–537.

[K1]   R. Kirby, *Stable homeomorphisms and the annulus conjecture*, Ann. of Math. **89** (1969) pp. 575–582 (also in [KS2, Annex A, pp. 291–298]).

[K2]   R. Kirby, *The topology of 4-manifolds*, Lecture Notes in Mathematics **1374** (Springer, Berlin–Heidelberg–New York, 1989).

[KS1]   R. Kirby and L. Siebenmann, *On the triangulation of manifolds and the Hauptvermutung*, Bull. Amer. Math. Soc. **75** (1969) pp. 742–749 (also in [KS2, Annex B, pp. 299–306]).

[KS2]   R. Kirby and L. Siebenmann, *Foundational Essays on Topological Manifolds, Smoothing and Triangulations*, Ann. of Math. Studies **88** (Princeton University Press, Princeton, 1977).

[Kis]   J. Kister, *Microbundles are fibre bundles*, Ann. of Math. (2) **80** (1964) pp. 190–199.

[KL]   N. Kuiper and R. Lashof, *Microbundles and bundles. I. Elementary theory*, Invent. Math. **1** (1966) pp. 1–17.

[LR1]   R. Lashof and M. Rothenberg, *Microbundles and smoothing*, Topology **3** (1965) pp. 357–388.

[LR2]   R. Lashof and M. Rothenberg, *Hauptvermutung for manifolds*, Proc 1967 Conference on the Topology of Manifolds, Prindle, Weber & Schmidt, Boston, Mass. (1968), pp. 81–105.

[M]   I. Madsen, *Higher torsion in SG and BSG*, Math. Z. **143** (1975) pp. 55–80.

[MM]   I. Madsen and R. Milgram, *The Classifying Spaces for Surgery and Cobordism of Manifolds*, Ann. of Math. Studies **92** (Princeton University Press, Princeton, 1979).

[Man]   C. Manolescu, *The Conley index, gauge theory, and triangulations*, J. Fixed Point Theory Appl. **13** (2013) no. 2, pp. 431–457.

[Mat]   T. Matumoto, *Triangulation of manifolds*, Algebraic and geometric topology (Proc. Sympos. Pure Math., Stanford Univ., Stanford, Calif., 1976), Part 2, pp. 3–6, Proc. Sympos. Pure Math., XXXII (Amer. Math. Soc., Providence, RI, 1978).

[May]   J. P. May, *Classifying spaces and fibrations*, Mem. Amer. Math. Soc. 1, no. **155** (Amer. Math. Soc., Providence, RI, 1975).

[McC]   M. C. McCord, *Classifying spaces and infinite symmetric products*, Trans. Amer. Math. Soc. **146** (1969) pp. 273–298.

[Mil]   R. Milgram, *Some remarks on the Kirby–Siebenmann class*, Algebraic topology and transformation groups (Göttingen, 1987), pp. 247–252, Lecture Notes in Math., **1361** (Springer, Berlin, 1988).

[Mi1]   J. Milnor, *On spaces having the homotopy type of CW-complex*, Trans. Amer. Math. Soc. **90** (1959) pp. 272–280.

[Mi2]   J. Milnor, *Two complexes which are homeomorphic but combinatorially distinct*, Ann. of Math. (2) **74** (1961) pp. 575–590.

[Mi3]   J. Milnor, *Microbundles and differential structures*, preprint http://faculty.tcu.edu/gfriedman/notes/milnor4.pdf

[Mi4]   J. Milnor, *Microbundles. I*, Topology **3** suppl. 1 (1964) 53–80.

[Mi5]   J. Milnor, *Lectures on the h-cobordism theorem* (Princeton University Press, Princeton, N.J., 1965).

[MK]   J. Milnor and M. Kervaire, *Bernoulli numbers, homotopy groups, and a theorem of Rohlin*, Proc. Internat. Congress Math. 1958, pp. 454–458 (Cambridge Univ. Press, New York, 1960).

[MiS]   J. Milnor and E. Spanier, *Two remarks on fiber homotopy type.* Pacific J. Math. **10** (1960) pp. 585–590.

[MS]   J. Milnor and J. Stasheff, *Characteristic classes.* Ann. of Math. Studies **76** (Princeton University Press, Princeton, 1974).

[Mo]   E. Moise, *Affine structures in 3-manifolds. V.* The triangulation theorem and Hauptvermutung, Ann. of Math. (2) **56** (1952) pp. 96–114.

[MT]   R. Mosher and M. Tangora, *Cohomology operations and applications in homotopy theory* (Harper & Row, New York–London, 1968, corrected reprint 2008).

[N1]   S. P. Novikov, *Homotopy equivalent smooth manifolds I*, Translations Amer Math. Soc. **48** (1965) pp. 271–396.

[N2]   S. P. Novikov, *On manifolds with free Abelian fundamental group and applications (Pontrjagin classes, smoothing, high–dimensional knots)*, Izvestiya AN SSSR, Ser. Math. **30** (1966) pp. 208–246.

[OV]   C. Okonek and A. Van de Ven, *Stable bundles and differentiable structures on certain elliptic surfaces*, Invent. Math. **86** (1986) no. 2, pp. 357–370.

[P]   Ch. Papakyriakopoulos, *A new proof for the invariance of the homology groups of a complex.* Bull. Soc. Math. Grèce **22** (1946) pp. 1–154.

[Q1]   F. Quinn, *Ends of maps and applications*, Bull. Amer. Math. Soc. (N.S.) **1** (1979) no. 1, 270–272.

[Q2]   F. Quinn, *Ends of maps, I*, Ann. of Math. **110** (1979) pp. 275–331.

[Q3]   F. Quinn, *Topological transversality holds in all dimensions*, Bull. Amer. Math. Soc. (N.S.) **18** (1988) no. 2, pp. 145–148.

[Q4]   F. Quinn, *A controlled-topology proof of the product structure theorem*, Geom. Dedicata **148** (2010) pp. 303–308.

[Rad]   T. Radó, *Über den Begriff der Riemannschen Fläche*, Acta Sci. Math. (Szeged) **2** (1925) pp. 101–121.

[Rand] D. Randall, *Equivalences to the triangulation conjecture*, Algebr. Geom. Topol. **2** (2002) pp. 1147–1154 (electronic).

[Ran]   A. Ranicki, *The Hauptvermutung book* (Kluwer, 1996).

[RW]   A. Ranicki and M. Weiss, *On the construction and topological invariance of the Pontryagin classes*, Geom. Dedicata **148** (2010) pp. 309–343.

[Ro]   V. A. Rokhlin, *New results in the theory of four-dimensional manifolds*, (Russian), Doklady Akad. Nauk SSSR (N.S.) **84** (1952) pp. 221–224.

[RoS]   V. Rokhlin and A. Švarc, *The combinatorial invariance of Pontrjagin classes*, (Russian) Dokl. Akad. Nauk SSSR (N.S.) **114** (1957) pp. 490–493.

[Rou]   C. P. Rourke, *The Hauptvermutung according to Casson and Sullivan*, In: The Hauptvermutung book (the item [Ran]), pp. 129–164.

[RS]   C. Rourke and B. Sanderson, *Introduction to piecewise-linear topology*, Reprint. Springer Study Edition (Springer-Verlag, Berlin-New York, 1982).

[Rud]   Yu. B. Rudyak, *On Thom spectra, Orientability, and Cobordism* (Springer, Berlin-Heidelberg-New York, 1998, corrected reprint 2008).

[Sav]   N. Saveliev, *Lectures on the topology of 3-manifolds. An introduction to the Casson invariant* (Walter de Gruyter & Co., Berlin, 1999).

[Sch]   M. Scharlemann, Transversality theories at dimension four, Invent. Math. **33** (1976) no. 1, pp. 1–14.

[Se]   J.-P. Serre, *Homologie singulière des espaces fibrés. Applications*, (French), Ann. of Math. (2) **54** (1951) pp. 425–505.

[S]   B. Sharma, *Topologically invariant integral characteristic classes*, Topology Appl. **21** (1985) pp. 135–146.

[SS]   B. Sharma and N. Singh, *Topological invariance of integral Pontrjagin classes mod p*, Topology Appl. **63** (1995) no. 1, pp. 59–67.

[Sieb1] L. Siebenmann, *The obstruction to finding a boundary for an open manifold of dimension greater then five*, Ph.D. Thesis, Princeton University (1965) 152 pp.

[Sieb2] L. Siebenmann, *Are nontriangulable manifolds triangulable?* 1970 Topology of Manifolds. Proc. Inst., Univ. of Georgia (Athens, GA, 1969), pp. 77–84.

[Sieb3] L. Siebenmann, *Approximating cellular maps by homeomorphisms*, Topology **11** (1972) pp. 271–294.

[Sieb4] L. Siebenmann, *Topological manifolds*, Proc ICM Nice (1970), Vol. 2, (Gauthier-Villars, Paris, 1971), pp. 133–163 (also in [KS2, Annex C, pp. 307–333]).

[Si]   N. Singh, *On topological and homotopy invariance of integral Pontrjagin classes modulo a prime p*, Topology Appl. **38** (1991) no. 3, pp. 225–235.

[Sma]   S. Smale, *Generalized Poincares conjecture in dimensions greater than four*, Ann. of Math. II. Ser. **74** (1961) pp. 391–406.

[Spa1]   E. Spanier, *Function spaces and duality*, Ann. of Math. (2) **70** (1959) pp. 338–378.

[Spa2]   E. Spanier, *Algebraic topology* (Springer-Verlag, New York-Berlin, 1981, corrected reprint).

[SW]   E. Spanier and J. Whitehead, *Duality in homotopy theory*, Mathematika **2** (1955) pp. 56–80.

[Spi]   M. Spivak, *Spaces satisfying Poincar duality*, Topology **6** (1967) pp. 77–101.

[Sta]   J. Stallings, *Lectures on polyhedral topology*, Tata Institute of Fundamental Research (Bombay 1967).

[St1]   N. Steenrod, *Cohomology operations*, Lectures by N. E. Steenrod written and revised by D. B. A. Epstein. Annals of Mathematics Studies, **50** (Princeton University Press, Princeton, N.J., 1962).

[St2]   N. Steenrod, *A convenient category of topological spaces*, Michigan Math. J. **14** (1967) pp. 133–152.

[Ste]   E. Steinitz, *Beiträge zur Analysis situs*, Sitz-Ber. Berlin Math. Ges. **7** (1908) pp. 29–49.

[Stong]   R. Stong, *Notes on cobordism theory* (Princeton University Press, Princeton, N.J., 1968).

[Sul1]   D. Sullivan, *On the Hauptvermutung for manifolds*, Bull. Amer. Math. Soc. **73** (1967) pp. 598–600.

[Sul2]   D. Sullivan, *Triangulating and smoothing homotopy equivalences and homeomorphisms. Geometric Topology Seminar Notes*, in: The Hauptvermutung book (the item [Ran]) pp. 69–103.

[ST]   D. Sullivan and N. Teleman, *An analytic proof of Novikov's theorem on rational Pontrjagin classes*, Inst. Hautes tudes Sci. Publ. Math. No. **58** (1983), 79–81 (1984).

[Sw]   R. Switzer, *Algebraic topology – homotopy and homology*, Die Grundlehren der mathematischen Wissenschaften, Band **212** (Springer-Verlag, New York-Heidelberg, 1975).

[Ta]   C. Taubes, *Gauge theory on asymptotically periodic 4-manifolds*, J. Differential Geom. **25** (1987) no. 3, 363–430.

[Ti]   H. Tietze, *Über die topologischen Invarianten mehrdimensionaler Mannigfaltigkeiten*, Monatsh. Math. Phys. **19** (1908) pp. 1–118.

[T]   R. Thom, *Les classes caractèristiques dc Pontrjagin des variètès triangulècs*, Symposium internacional de topologia algebraica (1958) pp. 54–67.

[W1]   C. T. C. Wall, *Determination of the cobordism ring*, Ann. of Math. (2) **72** (1960) pp. 292–311.

[W2]   C. T. C. Wall, *Poincaré complexes. I*, Ann. of Math. (2) **86** (1967) pp. 213–245.

[W3]   C. T. C. Wall, *On homotopy tori and the annulus theorem*, Bull. London Math. Soc. **1** (1969) pp. 95–97.

[W4]   C. T. C. Wall, *Surgery on compact manifolds*. Second edition. Edited and with a foreword by A. A. Ranicki. Mathematical Surveys and Monographs, **69** (Amer. Math. Soc., Providence, RI, 1999).

[Wash]   L. Washington, *Introduction to cyclotomic fields*. Second edition. Graduate Texts in Mathematics, **83** (Springer-Verlag, New York, 1997).

[Wh1]   G. Whitehead, *Generalized homology theories*, Trans. Amer. Math. Soc. **102** (1962) pp. 227–283.

[Wh2]   G. Whitehead, *Elements of homotopy theory*. Graduate Texts in Mathematics, **61** (Springer-Verlag, New York-Berlin, 1978).

[W]   J. Whitehead, *On C^1-Complexes*, Ann. of Math. **41** (1940) pp. 809–824.

[Wil]   R. Williamson, Jr, *Cobordism of combinatorial manifolds*, Ann. of Math. (2) **83** (1966) pp. 1–33.

[Wu]   W. T. Wu, *On Pontrjagin classes III*, Acta Math. Sinica, 4 (1954) pp. 323–346: English translation Amer. Math. Soc. Transl. Ser. 2, **11** (1959) pp. 155–172.

List of Symbols

Index

homotopy PL strucrure
  on a bundle, 20
  on a manifold, 19
homotopy triangulation, 19

induced bundle (fibration), 3
inessential map, 2
isotopy, 16

Kirby–Siebenmann class, 78
Kirby–Siebenmann invariant, 78
  universal, 78

lifting, 4
Local Contractibility Theorem, xxi

Main Theorem, xii, 35
microbundle, 2
morphism, 3
  over a map or space, 3

normal bordism, 46
  weak, 47
normal bordism class
  weak, 46
normal bundle, 22
normal invariant, 26, 46
normal morphism, 45
normally bordant morphisms, 46

oriented bordism, 53

PL $\mathbb{R}^n$-bundle, 9
PL atlas, 16
PL homeomorphism, 16
PL manifold, xi
PL morphism, 9
PL prestructure on a bundle, 17
PL structure on a bundle, 17
PL structure on a manifold, 16
plumbing, 31
pointed map, 1
pointed space, 1

Pontryagin class, 4
  modulo p, 84
  rational, 84
principal F-fibration, 4
Product Structure Theorem, xxi, 28,
  91, 92
proper map, 2

reducibility, 25
reducible map, 25
Reduction Theorem, 35
Rokhlin Signature Theorem, xxii, 30

signature, 4
splittable map, 59
splitting, 60
stable duality, 38
stable homotopy class, 37
Steenrod–Thom homorphism, 53
Stiefel–Whitney class, 4
surgery exact sequence, 27, 50

tame map, 92
tangent bundle, 22
Thom space
  of a spherical fibration, 24
  of an $\mathbb{R}^n$-bundle, 24
topological $\mathbb{R}^n$-bundle, 7
topological $\mathbb{R}^n$-morphism, 7
topologically invariant class, 85
total space, 3
trivial F-bundle, 2
trivial element of $\mathcal{S}_{PL}(M)$, 19
trivial element of $\mathcal{T}_{PL}(M)$, 16
trivial principal F-fibration, 4
trnsversally splittable map, 60

universal property, 8–10

vertically homotopic liftings, 4

Whitney sum, 11

Printed in the United States
By Bookmasters